Eclipse Almanac 2051 To 2060

Ten Years of Solar and Lunar Eclipses

Color Edition

Fred Espenak

Edition 1.0
October 2020

Eclipse Almanac 2051 To 2060 – Color Edition

Ten Years of Solar and Lunar Eclipses

Astropixels Publishing
P.O. Box 16197
Portal, AZ 85632

www.astropixels.com/pubs

This book may be ordered at: *www.astropixels.com/pubs/EclipseAlmanac.html*

More about solar and lunar eclipses from 2051 To 2060 can be found at EclipseWise:

www.astropixels.com/news/EclipseAlmanac.html

Astropixels Publication Number: AP032

First Edition (Version 1.0a)

ISBN 978-1-941983-32-4

Printed in the United States of America

Front Cover: Drawing of the Total Lunar Eclipse of 1550 October 6, ***Astronomicum Caesareum*** (Peter Apian, 1540)

Back Cover: Portrait of Fred Espenak (Copyright ©2018 by Fred Espenak).

Preface

The ***Eclipse Almanac*** contains maps and diagrams of every solar and lunar eclipse over a ten-year period. This permits the reader to look ahead and easily determine when and where each of these events will be seen. Particular details about each eclipse are included as well as a 25-year table looking further into the future.

Section 1 covers solar eclipses, while Section 2 is devoted to lunar eclipses. Brief explanations are given for the different types of eclipses, and descriptions of the visual appearance of each one is included. Section 3 tabulates the date and time of the Moon's phases over a decade. New Moon and Full Moon phases that coincide with solar and lunar eclipses, respectively, are identified.

The ***Eclipse Almanac*** series consists of five volumes. Each one covering a single decade as follows:

1) Eclipse Almanac 2021 to 2030   4) Eclipse Almanac 2051 to 2060
2) Eclipse Almanac 2031 to 2040   5) Eclipse Almanac 2061 to 2070
3) Eclipse Almanac 2041 to 2050

Each volume is available in a color edition as well as a more economical black and white edition.

All times listed in the ***Eclipse Almanac*** are in Universal Time (UT1). This is the modern replacement of Greenwich Mean Time, which was based on mean solar time from Greenwich, England. In comparison, Universal Time is based on Earth's rotation with respect to distant quasars.

For North Americans, the conversion from UT1 to local time is as follows:

Atlantic Standard Time (AST)	= UT1 - 4 hours
Eastern Standard Time (EST)	= UT1 - 5 hours
Central Standard Time (CST)	= UT1 - 6 hours
Mountain Standard Time (MST)	= UT1 - 7 hours
Pacific Standard Time (PST)	= UT1 - 8 hours

If Daylight Saving Time is in effect in the time zone, you must ADD one hour to the standard time.

The conversion of Universal Time to other time zones is easily found on the Internet.

For more information about solar and lunar eclipses from 2021 to 2070, visit *EclipseWise.com* at:

www.astropixels.com/news/EclipseAlmanac.html

Table of Contents

Central Solar Eclipses from 2051 to 2060

The path of every central solar eclipse from 2051 to 2060 is plotted on a world map. The central paths of total eclipses are shaded blue, while annular eclipses are shaded red. For hybrid eclipses, part of the path is shaded blue (total), and part is shaded red (annular). Major cities are plotted as black dots, scaled by population size. (©2020 F. Espenak)

Photo 1–1 The "Diamond Ring Effect" is seen just before totality begins. Total Solar Eclipse of 2018 Jul 03. ©2018 F. Espenak

Section 1: Solar Eclipses

Introduction

The Moon orbits Earth once every 29.5306 days with respect to the Sun. Over the course of its orbit, the Moon's changing position relative to the Sun results in its familiar phases: New Moon > First Quarter > Full Moon > Last Quarter > New Moon. The New Moon phase is the only one not visible because the illuminated side of the Moon then points away from Earth.

The orbit of the Moon is tilted about 5.1° to Earth's orbit around the Sun. The points where the two orbits appear to cross are called the nodes. When the New Moon occurs near one of these nodes, the Moon's shadow falls on a portion of Earth and a solar eclipse is visible from that region.

The Moon's shadow is composed of three cone-shaped components. The outer or penumbral shadow is a zone where the Sun's rays are partially blocked. Nested within the penumbra is the umbral shadow — a region where direct rays from the Sun are completely blocked. The conical umbra tapers to a point beyond which extends an expanding cone called the antumbra. From within this third shadow, the Moon appears smaller than the Sun and is seen in silhouette against the solar disk.

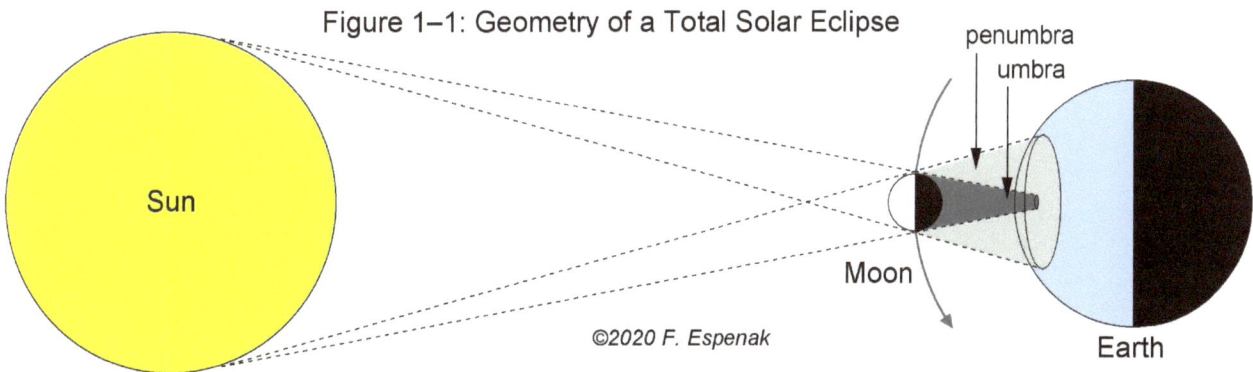

Figure 1–1 illustrates the geometry of a total solar eclipse. A partial eclipse is visible from within the large penumbral shadow, while the total eclipse is only seen from the much smaller umbral shadow.

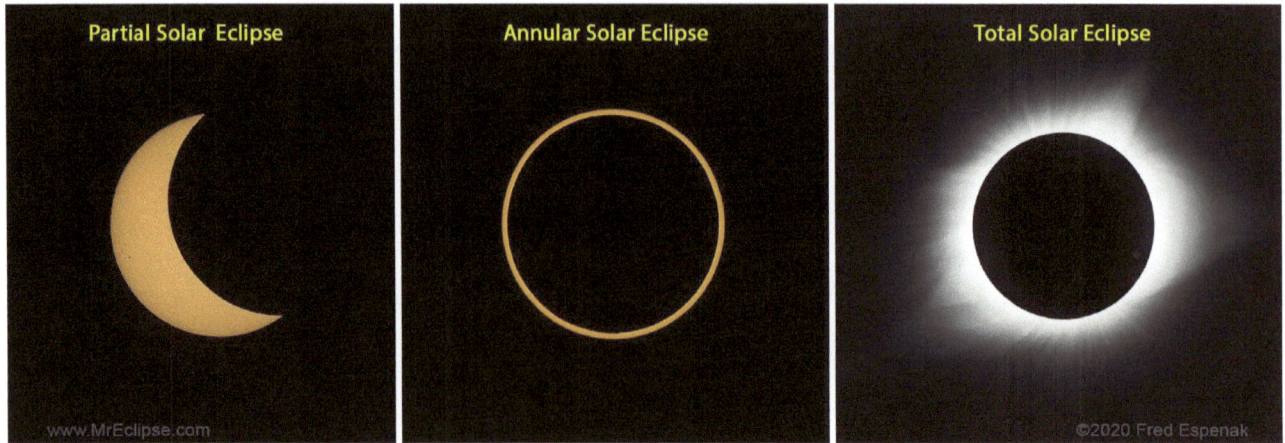

Photo 1–2 Examples of a partial, annular, and total solar eclipse. ©2020 F. Espenak

Types of Solar Eclipses

There are four types of solar eclipses:

1. **Total Solar Eclipse** — The Moon's penumbral and umbral shadows traverse Earth. The Moon's antumbral shadow extends beyond Earth's surface. The Moon appears larger than the Sun and completely covers the solar disk when viewed from within the umbral shadow. A partial eclipse is seen within the penumbral shadow (Figure 1–1).
2. **Annular Solar Eclipse** — The Moon's penumbral and antumbral shadows traverse Earth. The Moon's umbral shadow completely misses Earth. The Moon's disk appears smaller than the Sun so a bright ring of the Sun's disk surrounds the Moon when viewed from within the antumbral shadow. A partial eclipse is seen within the penumbral shadow (Figure 1–2).
4. **Hybrid Solar Eclipse** — The Moon's penumbral, umbral and antumbral shadows all traverse parts of Earth. The curvature of Earth's surface brings some regions into the umbra and others into the antumbra. The eclipse is total within the umbra and annular within the antumbra. A partial eclipse is seen within the penumbral shadow. Hybrid eclipses are also known as annular-total eclipses.
4. **Partial Solar Eclipse** — The Moon's penumbral shadow traverses Earth while the umbral and antumbral shadows miss Earth. A portion of the Sun's disk is obscured from within the penumbra.

Figure 1–2 illustrates the geometry of an annular solar eclipse. A partial eclipse is visible from within the large penumbral shadow, while the annular eclipse is confined to the much smaller antumbral shadow.

Annular, total and hybrid eclipses are sometimes referred to as *central* eclipses[1]. However, on rare occasions it is possible to have an annular or total eclipse that is non-central. This occurs in Earth's polar regions when only the edge of the umbral or antumbral falls on Earth while the shadow axis misses the planet entirely. The eclipse of 2043 April 09 is a non-central total solar eclipse.

---

[1] A central eclipse is one in which the central axis of the Moon's umbral/antumbral shadow intersects with Earth's surface.

Photo 1–3 Partial solar eclipse of 2014 Oct 23. ©2014 F. Espenak

Visual Appearance of Partial Solar Eclipses

When the Moon's penumbral shadow strikes Earth, a partial eclipse of the Sun is visible from that region. The apparent motion of the Moon with respect to the Sun is gradual — the partial phases can two hours or more. During this time, the Moon's dark limb slowly creeps across the Sun's disk.

Partial eclipses cannot be viewed with the unprotected eye because the Sun is still extremely bright. Special techniques are needed to safely view the eclipse (for example, special eclipse filters or glasses). Even when a partial eclipse reaches its maximum phase, the sky and landscape remain bright. But careful inspection of dappled sunlight beneath a leafy tree will reveal multiple images of the eclipse. The gaps between the tree leaves act like pinhole cameras and project images of the crescent Sun onto the ground.

The Moon's penumbral shadow is typically 4200 to 4500 miles (6700 to 7300 kilometers) in diameter and can cover a significant fraction of the daytime hemisphere of Earth. Consequently, partial eclipses are visible from large geographic regions as the penumbra sweeps across Earth's surface.

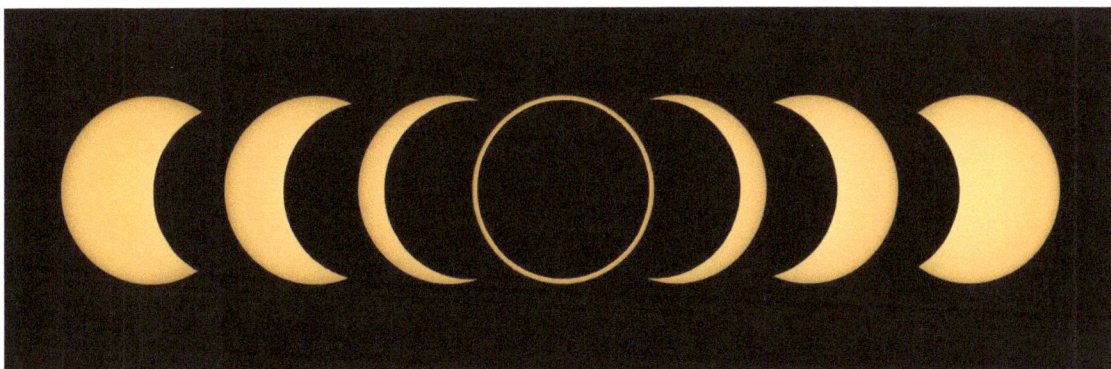

Photo 1–4 shows various phases of the Annular solar eclipse of 2005 Oct 03. ©2005 F. Espenak

Visual Appearance of Annular Solar Eclipses

During an annular eclipse, the Moon's penumbral and antumbral shadows sweep across Earth. Compared to the penumbra, the antumbra is much smaller and has a maximum diameter of 232 miles (374 kilometers). Because of this, the antumbra covers a tiny fraction of Earth's surface.

A partial eclipse is visible within the penumbral shadow, but only observers located in the much narrower track of the antumbra will see an annular eclipse. For this reason, the antumbra's trajectory across Earth is called the path of annularity.

All annular eclipses begin with a series of partial phases lasting an hour or more. At the peak of the eclipse, the Moon's disk can be seen in complete silhouette against the Sun. The remaining solar photosphere appears as an intensely bright ring of light surrounding the Moon. The annular phase can last a maximum of 12 ½ minutes but is more typically 3 to 5 minutes in length. After annularity, another series of partial phases occur as the Moon gradually uncovers the Sun.

Special precautions must be used to watch an annular eclipse (just like partial eclipses). Even during the annular phase, the Sun is dangerously bright and cannot be viewed without a solar filter. The landscape and sky remain bright throughout the eclipse, giving little indication of the celestial event in progress.

Photo 1–5 The Sun's corona is only visible to the naked eye for a few minutes during a total eclipse of the Sun. This is an HDR composite image of the total solar eclipse of August 21, 2017 from Casper, Wyoming. © 2017 F. Espenak, www.MrEclipse.com

Visual Appearance of Total Solar Eclipses

During a total eclipse, the Moon's penumbral and umbral shadows fall upon Earth. The umbra has a maximum diameter of 170 miles (273 kilometers). The narrow track traced out as the umbra sweeps across Earth's surface is called the path of totality. The Sun is completely obscured by the Moon from within this zone. The total phase can last up to 7 ½ minutes but is more typically 2 to 3 minutes in length.

Total eclipses all begin and end with a series of partial phases lasting an hour or more. But this is where the resemblance to partial and annular eclipses ends — the total phase is the most spectacular astronomical event visible to the naked eye. During totality, the Sun's outer atmosphere — the solar corona — appears as a gossamer halo surrounding the Moon, and bright stars and planets are visible.

The eclipse takes on a unique character about five minutes before the total phase commences. The Sun has a foreboding quality and casts abnormally sharp shadows. The approaching lunar umbra darkens the western sky and the air temperature is noticeably cooler. A minute before totality, pale shadow bands[2] ripple across the ground.

The ambient light grows feeble even though the crescent Sun is still too bright to see. In the final seconds, the Sun's corona emerges from the glare as the solar crescent shrinks to a brilliant jewel. This celestial diamond ring lingers for a moment before the sunlight is extinguished and totality begins.

The Sun's glorious corona is now displayed to full advantage in the darkened sky. Standing within the Moon's umbra affords the rare and unprecedented opportunity to gaze directly at the glowing million-degree plasma surrounding our star. Twisted, tortured, and constrained by the Sun's enormous magnetic fields, the solar corona is revealed to the naked eye only during the brief seconds when the Moon completely blocks the brilliant disk of the Sun. An eerie twilight bathes the landscape and the colors of dusk surround the horizon.

Minutes race by like seconds. Suddenly, a sparkling bead of sunlight reappears along one edge of the Moon and quickly grows to blindingly bright proportions. Daylight returns as the corona fades and the totality ends. Another hour of partial phases remains before the eclipse is over.

While filters are required for viewing the partial phases, they must be removed for totality. The total phase is the only time it is completely safe to look directly at the Sun without protection. In fact, the total phase is not even visible through solar filters because the Sun's corona is a million times fainter than the photosphere[3].

Figure 1–3: Solar Eclipse Contacts

Figure 1–3 illustrates the four contacts for annular and total solar eclipses. The arrows indicate the contact point of the Moon with the Sun's limb in each diagram.

Solar Eclipse Contacts

During the course of a solar eclipse, the instants when the edge of the Moon's disk becomes tangent to the Sun's disk are known as eclipse contacts. They mark various stages or phases of a solar eclipse (Figure 1–3).

Partial solar eclipses have two primary contacts.

First Contact (C1) — Partial Eclipse Begins (Instant of first exterior tangency of the Moon with the Sun)
Fourth Contact (C4) — Partial Eclipse Ends (Instant of last exterior tangency of the Moon with the Sun)

---

[2] Shadow bands are caused by the thin solar crescent illuminating Earth's atmosphere before and after totality.
[3] The photosphere is the visible surface of the Sun's disk.

Central solar eclipses (total, annular or hybrid) have four primary contacts. Contacts C2 and C3 mark the instants when the Moon's disk is first and last internally tangent to the Sun. These are the times when the annular or total phase of the eclipse begins and ends, respectively.

First Contact (C1) — Partial Eclipse Begins (Instant of first exterior tangency of the Moon with the Sun)
Second Contact (C2) — Total (or Annular) Eclipse Begins (Instant of first interior tangency of Moon with Sun)
Third Contact (C3) — Total (or Annular) Eclipse Ends (Instant of last interior tangency of Moon with Sun)
Fourth Contact (C4) — Partial Eclipse Ends (Instant of last exterior tangency of the Moon with the Sun)

Photo 1–6 Sequence of partial phases, Diamond Rings, and Totality. Total Solar Eclipse of 2017 Aug 21. ©2017 F. Espenak

Explanation of Global Solar Eclipse Maps

There are 23 eclipses of the Sun during the period 2051 to 2060. A global map is included for each eclipse.

The geographic visibility of each eclipse is illustrated with an orthographic projection map of Earth showing the path of the Moon's penumbral and umbral/antumbral shadows with respect to the continental coastlines, political boundaries (circa 2016), and major cities (represented by black dots). North is up and the daylight terminator is drawn for the instant of greatest eclipse. An asterisk marks the sub-solar point where the Sun appears directly overhead at that time. The salient features of the eclipse maps are identified in the key diagram on page 11.

The limits of the Moon's penumbral shadow delineate the region where a partial solar eclipse is visible. This irregular or saddle shaped region often covers more than half the daylight hemisphere of Earth and consists of several distinct zones or limits. At the northern and/or southern boundaries lie the limits of the penumbra's path. Partial eclipses have only one of these limits, as do central eclipses when the Moon's shadow axis falls no closer than about 0.45 radii from Earth's center. Great loops at the western and eastern extremes of the penumbra's path identify the areas where the eclipse begins/ends at sunrise and sunset, respectively. If the penumbra has both a northern and southern limit, the rising and setting curves form two separate, closed loops (e.g., 2024 Apr 08). Otherwise, the curves are connected in a distorted figure eight (e.g., 2021 Dec 04). Bisecting the *eclipse begins/ends at sunrise and sunset* loops is the curve of maximum eclipse at sunrise (western loop) and sunset (eastern loop).

The eclipse magnitude is defined as the fraction of the Sun's diameter occulted by the Moon. A curve of constant eclipse magnitude delineates the points where the local magnitude at maximum eclipse is equal to a constant value. The maps include *curves of constant eclipse magnitude* for values of 0.2, 0.4, 0.6, and 0.8. These curves run exclusively between the curves of maximum eclipse at sunrise and sunset. They are approximately parallel to the northern/southern penumbral limits and the umbral/antumbral paths of central eclipses. The northern and southern limits of the penumbra may be thought of as curves of eclipse magnitude of 0.0. For total eclipses, the northern and southern limits of the umbra are curves of eclipse magnitude of 1.0.

Greatest eclipse is the instant when the axis of the Moon's shadow cone passes closest to Earth's center. The point on Earth's surface intersected by the axis of the Moon's shadow cone at greatest eclipse is marked by an asterisk. For partial eclipses, the shadow axis misses Earth entirely, so the point of greatest eclipse lies on the day/night terminator and the Sun appears on the horizon.

Parameters for the eclipse appears to the right of each map. They include the time of greatest eclipse (Universal Time[4] or UT1), the eclipse magnitude and obscuration, gamma[5], Saros series, and node (ascending or descending).

Photo 1–7 Time Sequence of the Total Solar Eclipse of 2017 Aug 21. ©2017 F. Espenak

---

[4] Universal Time (UT1) is the modern-day replacement for Greenwich Mean Time.
[5] Gamma is the distance of the Moon's shadow cone axis from Earth's center at greatest eclipse (in Earth equatorial radii).

Global Solar Eclipse Maps

Key to Global Solar Eclipse Maps

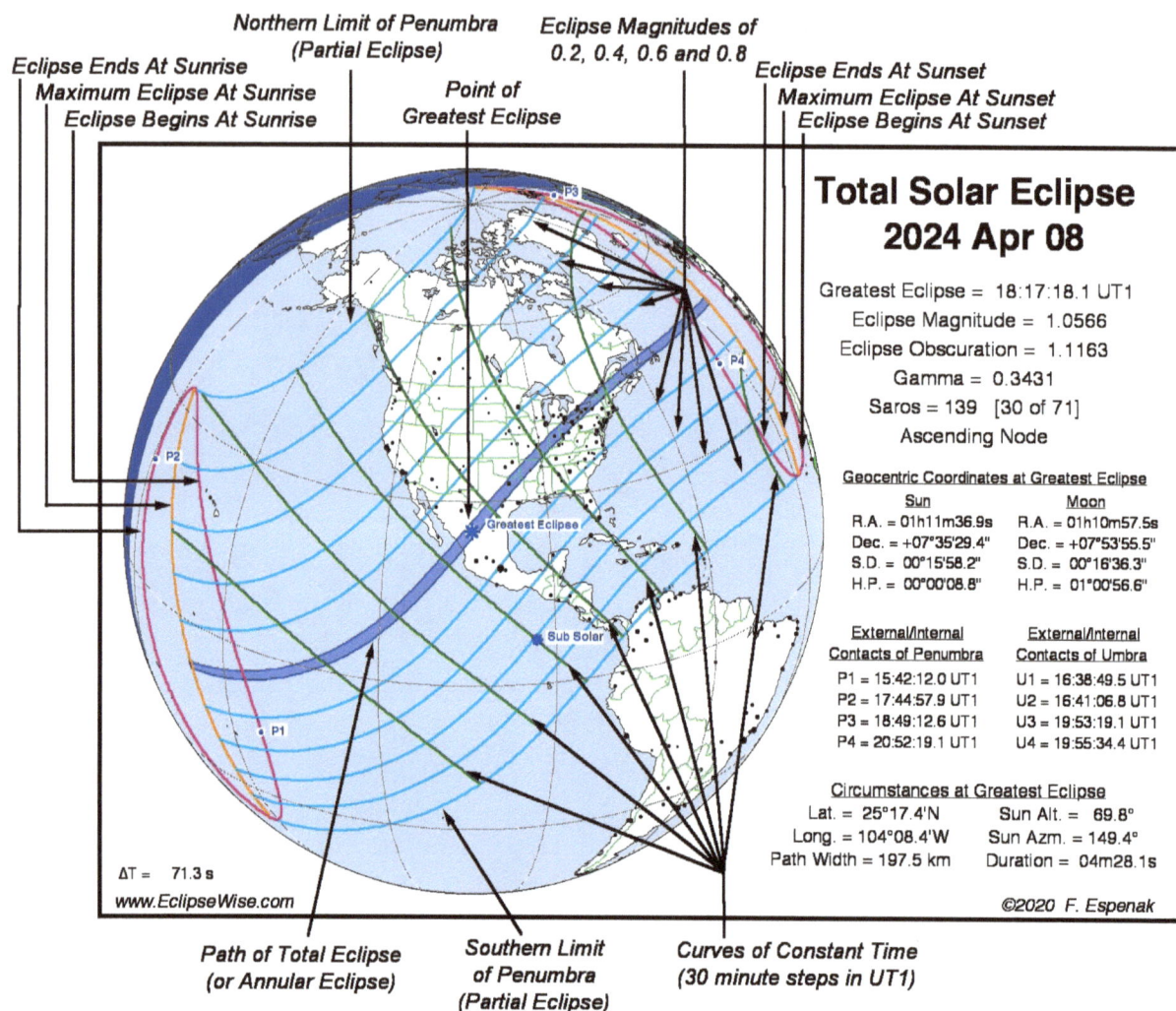

Diagram labels (clockwise from top):
Eclipse Ends At Sunrise / Maximum Eclipse At Sunrise / Eclipse Begins At Sunrise
Northern Limit of Penumbra (Partial Eclipse)
Point of Greatest Eclipse
Eclipse Magnitudes of 0.2, 0.4, 0.6 and 0.8
Eclipse Ends At Sunset / Maximum Eclipse At Sunset / Eclipse Begins At Sunset
Path of Total Eclipse (or Annular Eclipse)
Southern Limit of Penumbra (Partial Eclipse)
Curves of Constant Time (30 minute steps in UT1)

Total Solar Eclipse 2024 Apr 08

Greatest Eclipse = 18:17:18.1 UT1
Eclipse Magnitude = 1.0566
Eclipse Obscuration = 1.1163
Gamma = 0.3431
Saros = 139 [30 of 71]
Ascending Node

Geocentric Coordinates at Greatest Eclipse

Sun	Moon
R.A. = 01h11m36.9s	R.A. = 01h10m57.5s
Dec. = +07°35'29.4"	Dec. = +07°53'55.5"
S.D. = 00°15'58.2"	S.D. = 00°16'36.3"
H.P. = 00°00'08.8"	H.P. = 01°00'56.6"

External/Internal Contacts of Penumbra	External/Internal Contacts of Umbra
P1 = 15:42:12.0 UT1	U1 = 16:38:49.5 UT1
P2 = 17:44:57.9 UT1	U2 = 16:41:06.8 UT1
P3 = 18:49:12.6 UT1	U3 = 19:53:19.1 UT1
P4 = 20:52:19.1 UT1	U4 = 19:55:34.4 UT1

Circumstances at Greatest Eclipse

Lat. = 25°17.4'N	Sun Alt. = 69.8°
Long. = 104°08.4'W	Sun Azm. = 149.4°
Path Width = 197.5 km	Duration = 04m28.1s

ΔT = 71.3 s
www.EclipseWise.com
©2020 F. Espenak

Explanation of Terms Used in Global Solar Eclipse Maps

Greatest Eclipse – The instant when the distance between the axis of the Moon's shadow cone and the center of Earth reaches a minimum (in Universal Time[6] or UT1).

Eclipse Magnitude – The fraction of the Sun's diameter occulted by the Moon at the instant of greatest eclipse (for total and annular eclipses this value is the ratio of diameters of the Moon and the Sun).

Eclipse Obscuration – The fraction of the Sun's area occulted by the Moon at the instant of greatest eclipse.

Gamma – The minimum distance from the lunar shadow axis to the center of Earth (units of Earth equatorial radii).

Saros Series – The Saros series that the eclipse belongs to. The numbers in "[ ]" are the eclipse's sequential position and the number of eclipses in the Saros series.

Node – The orbital node near which the eclipse takes place (Ascending Node or Descending Node).

Geocentric Coordinates of the Sun and the Moon at Greatest Eclipse

      R.A. – Right Ascension   **S.D.** – Semi-Diameter (i.e. - radius)

      Dec. – Declination   **H.P.** – Horizontal Parallax

External/Internal Contacts of Penumbra and Umbra – Instants when each shadow enters or exits the surface of Earth (Penumbral contacts shown on map as: P1, P2, P3, and P4; umbral contacts are located at the ends of the central path). All times are in Universal Time (UT1)

Circumstances of Greatest Eclipse – Geographic location (Lat., Long.) of the shadow axis, the altitude and azimuth of the Sun, width of the central path, and the central duration of totality or annularity.

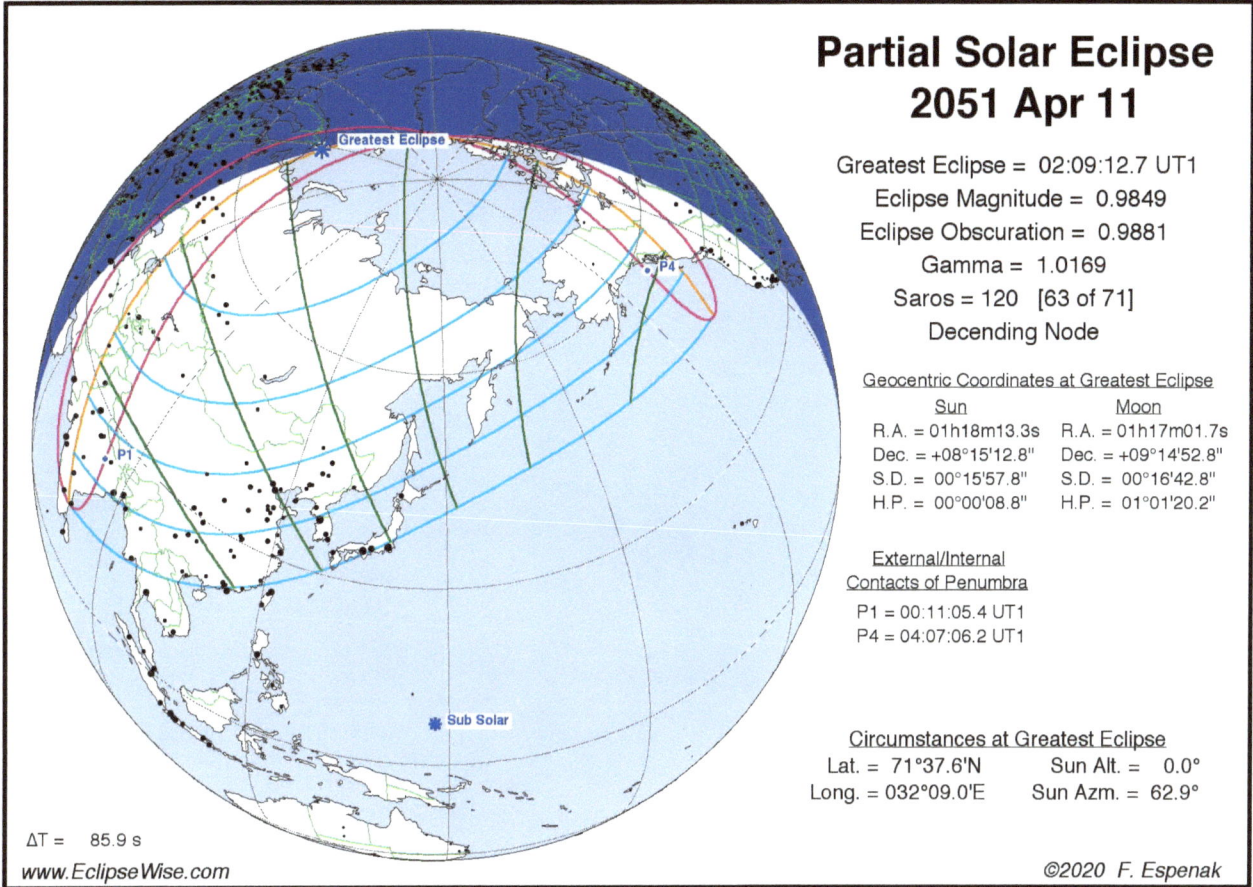

Partial Solar Eclipse
2051 Apr 11

Greatest Eclipse = 02:09:12.7 UT1

Eclipse Magnitude = 0.9849

Eclipse Obscuration = 0.9881

Gamma = 1.0169

Saros = 120 [63 of 71]

Decending Node

Geocentric Coordinates at Greatest Eclipse

	Sun	Moon
R.A. =	01h18m13.3s	R.A. = 01h17m01.7s
Dec. =	+08°15'12.8"	Dec. = +09°14'52.8"
S.D. =	00°15'57.8"	S.D. = 00°16'42.8"
H.P. =	00°00'08.8"	H.P. = 01°01'20.2"

External/Internal
Contacts of Penumbra

P1 = 00:11:05.4 UT1
P4 = 04:07:06.2 UT1

Circumstances at Greatest Eclipse

Lat. = 71°37.6'N	Sun Alt. = 0.0°
Long. = 032°09.0'E	Sun Azm. = 62.9°

ΔT = 85.9 s

www.EclipseWise.com

©2020 F. Espenak

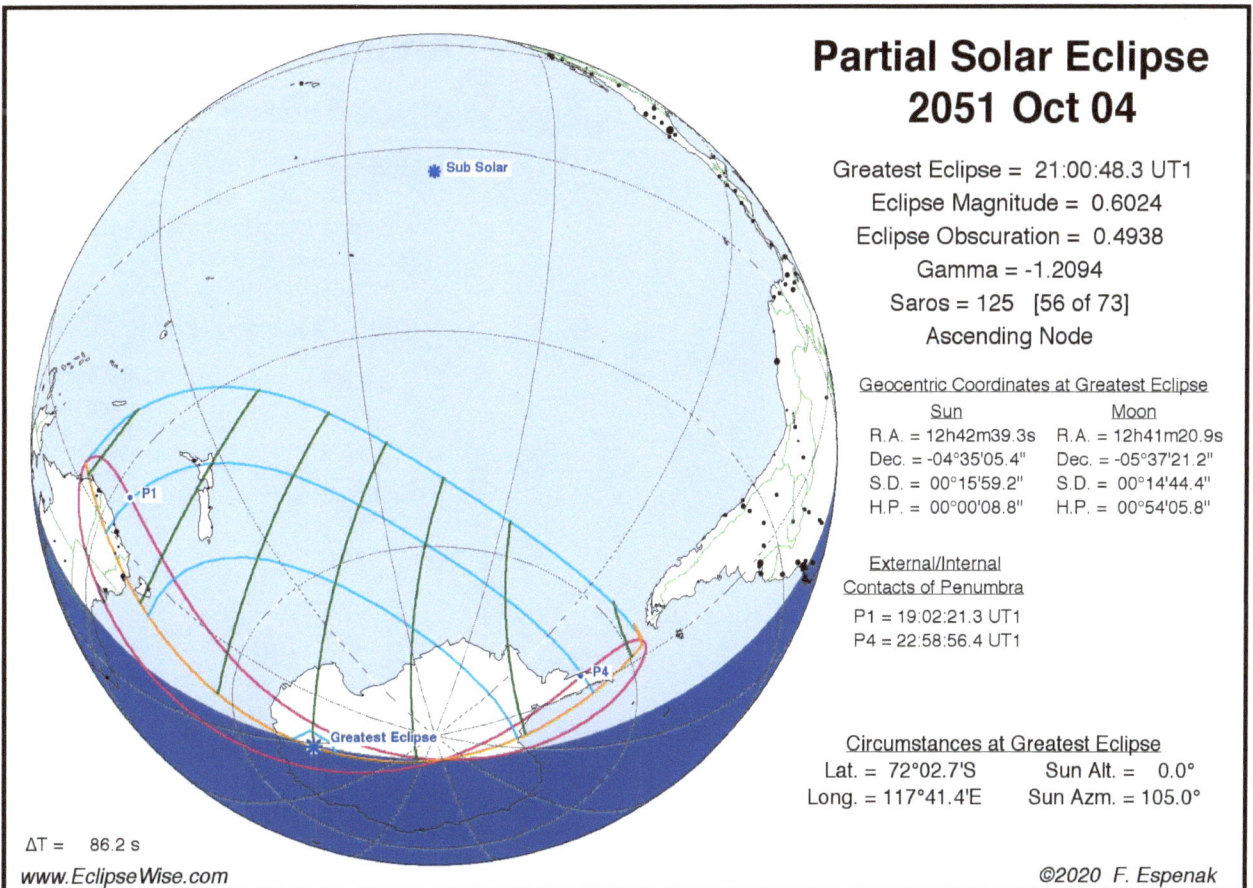

Partial Solar Eclipse
2051 Oct 04

Greatest Eclipse = 21:00:48.3 UT1

Eclipse Magnitude = 0.6024

Eclipse Obscuration = 0.4938

Gamma = -1.2094

Saros = 125 [56 of 73]

Ascending Node

Geocentric Coordinates at Greatest Eclipse

	Sun	Moon
R.A. =	12h42m39.3s	R.A. = 12h41m20.9s
Dec. =	-04°35'05.4"	Dec. = -05°37'21.2"
S.D. =	00°15'59.2"	S.D. = 00°14'44.4"
H.P. =	00°00'08.8"	H.P. = 00°54'05.8"

External/Internal
Contacts of Penumbra

P1 = 19:02:21.3 UT1
P4 = 22:58:56.4 UT1

Circumstances at Greatest Eclipse

Lat. = 72°02.7'S	Sun Alt. = 0.0°
Long. = 117°41.4'E	Sun Azm. = 105.0°

ΔT = 86.2 s

www.EclipseWise.com

©2020 F. Espenak

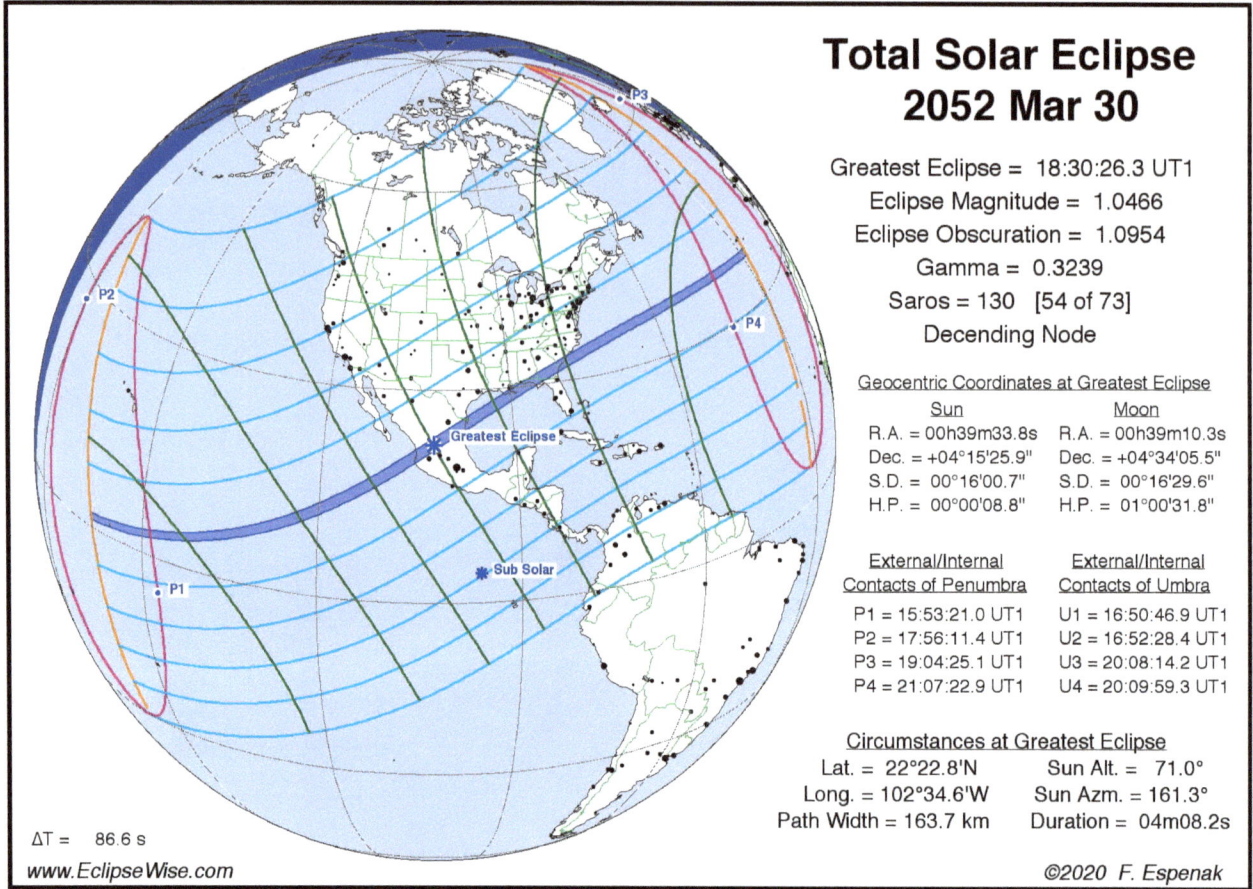

Total Solar Eclipse
2052 Mar 30

Greatest Eclipse = 18:30:26.3 UT1

Eclipse Magnitude = 1.0466

Eclipse Obscuration = 1.0954

Gamma = 0.3239

Saros = 130 [54 of 73]

Decending Node

Geocentric Coordinates at Greatest Eclipse

Sun	Moon
R.A. = 00h39m33.8s	R.A. = 00h39m10.3s
Dec. = +04°15'25.9"	Dec. = +04°34'05.5"
S.D. = 00°16'00.7"	S.D. = 00°16'29.6"
H.P. = 00°00'08.8"	H.P. = 01°00'31.8"

External/Internal Contacts of Penumbra	External/Internal Contacts of Umbra
P1 = 15:53:21.0 UT1	U1 = 16:50:46.9 UT1
P2 = 17:56:11.4 UT1	U2 = 16:52:28.4 UT1
P3 = 19:04:25.1 UT1	U3 = 20:08:14.2 UT1
P4 = 21:07:22.9 UT1	U4 = 20:09:59.3 UT1

Circumstances at Greatest Eclipse

Lat. = 22°22.8'N	Sun Alt. = 71.0°
Long. = 102°34.6'W	Sun Azm. = 161.3°
Path Width = 163.7 km	Duration = 04m08.2s

ΔT =   86.6 s

www.EclipseWise.com

©2020 F. Espenak

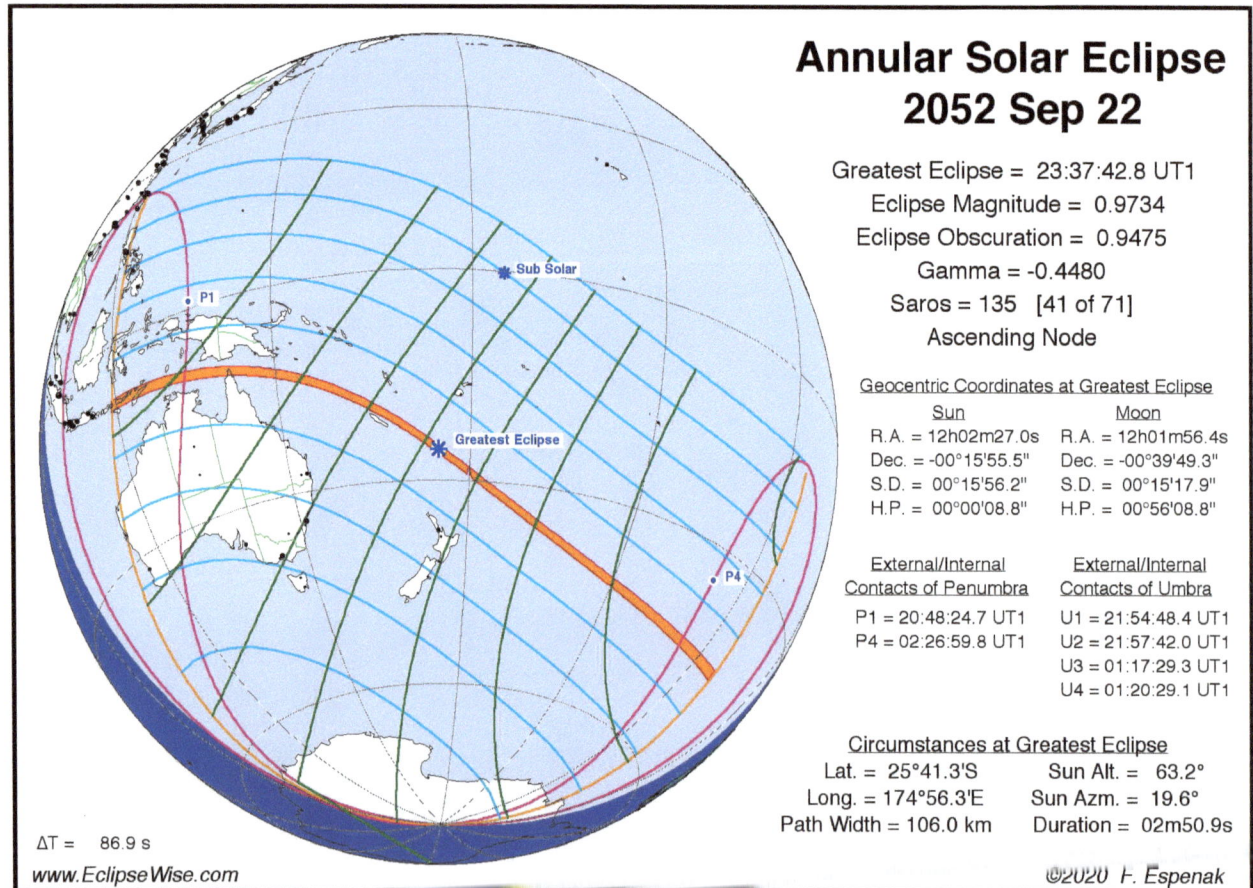

Annular Solar Eclipse
2052 Sep 22

Greatest Eclipse = 23:37:42.8 UT1

Eclipse Magnitude = 0.9734

Eclipse Obscuration = 0.9475

Gamma = -0.4480

Saros = 135 [41 of 71]

Ascending Node

Geocentric Coordinates at Greatest Eclipse

Sun	Moon
R.A. = 12h02m27.0s	R.A. = 12h01m56.4s
Dec. = -00°15'55.5"	Dec. = -00°39'49.3"
S.D. = 00°15'56.2"	S.D. = 00°15'17.9"
H.P. = 00°00'08.8"	H.P. = 00°56'08.8"

External/Internal Contacts of Penumbra	External/Internal Contacts of Umbra
P1 = 20:48:24.7 UT1	U1 = 21:54:48.4 UT1
P4 = 02:26:59.8 UT1	U2 = 21:57:42.0 UT1
	U3 = 01:17:29.3 UT1
	U4 = 01:20:29.1 UT1

Circumstances at Greatest Eclipse

Lat. = 25°41.3'S	Sun Alt. = 63.2°
Long. = 174°56.3'E	Sun Azm. = 19.6°
Path Width = 106.0 km	Duration = 02m50.9s

ΔT =   86.9 s

www.EclipseWise.com

©2020 F. Espenak

Annular Solar Eclipse
2053 Mar 20

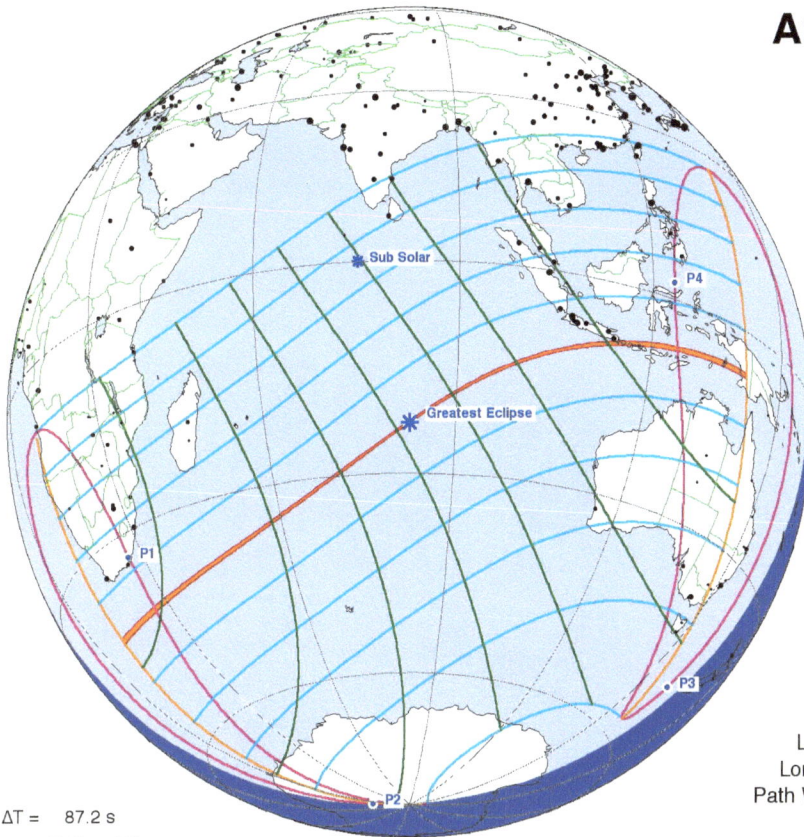

Greatest Eclipse = 07:06:52.2 UT1

Eclipse Magnitude = 0.9919

Eclipse Obscuration = 0.9839

Gamma = -0.4089

Saros = 140 [31 of 71]

Decending Node

Geocentric Coordinates at Greatest Eclipse

Sun	Moon
R.A. = 00h00m30.3s	R.A. = 00h00m59.0s
Dec. = +00°03'17.2"	Dec. = -00°19'05.6"
S.D. = 00°16'03.6"	S.D. = 00°15'41.9"
H.P. = 00°00'08.8"	H.P. = 00°57'37.0"

External/Internal Contacts of Penumbra	External/Internal Contacts of Umbra
P1 = 04:21:39.0 UT1	U1 = 05:25:32.6 UT1
P2 = 06:47:48.1 UT1	U2 = 05:27:09.5 UT1
P3 = 07:26:21.6 UT1	U3 = 08:46:47.8 UT1
P4 = 09:52:05.7 UT1	U4 = 08:48:18.6 UT1

Circumstances at Greatest Eclipse

Lat. = 23°00.9'S	Sun Alt. = 65.7°
Long. = 082°54.4'E	Sun Azm. = 340.8°
Path Width = 31.1 km	Duration = 00m49.5s

ΔT =   87.2 s

www.EclipseWise.com

©2020 F. Espenak

Total Solar Eclipse
2053 Sep 12

Greatest Eclipse = 09:32:41.4 UT1

Eclipse Magnitude = 1.0328

Eclipse Obscuration = 1.0668

Gamma = 0.3140

Saros = 145 [24 of 77]

Ascending Node

Geocentric Coordinates at Greatest Eclipse

Sun	Moon
R.A. = 11h23m36.1s	R.A. = 11h23m58.5s
Dec. = +03°55'14.2"	Dec. = +04°12'57.2"
S.D. = 00°15'53.4"	S.D. = 00°16'09.4"
H.P. = 00°00'08.7"	H.P. = 00°59'17.8"

External/Internal Contacts of Penumbra	External/Internal Contacts of Umbra
P1 = 06:51:44.1 UT1	U1 = 07:51:00.0 UT1
P2 = 08:57:26.9 UT1	U2 = 07:52:00.9 UT1
P3 = 10:08:11.6 UT1	U3 = 11:13:32.6 UT1
P4 = 12:13:48.9 UT1	U4 = 11:14:28.5 UT1

Circumstances at Greatest Eclipse

Lat. = 21°28.1'N	Sun Alt. = 71.6°
Long. = 041°40.1'E	Sun Azm. = 198.6°
Path Width = 116.4 km	Duration = 03m04.1s

ΔT =   87.5 s

www.EclipseWise.com

©2020 F. Espenak

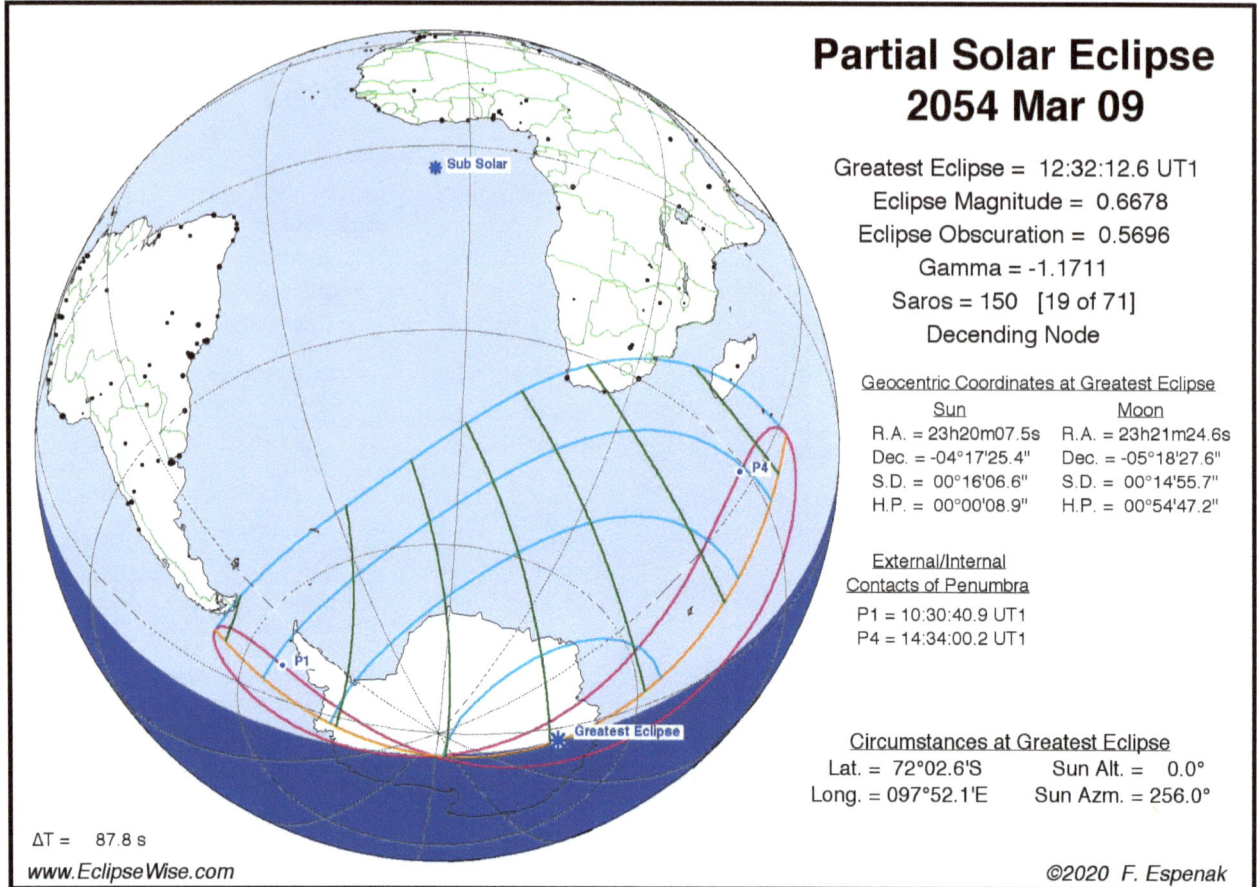

Partial Solar Eclipse
2054 Mar 09

Greatest Eclipse = 12:32:12.6 UT1

Eclipse Magnitude = 0.6678

Eclipse Obscuration = 0.5696

Gamma = -1.1711

Saros = 150   [19 of 71]

Decending Node

Geocentric Coordinates at Greatest Eclipse

	Sun	Moon
R.A. =	23h20m07.5s	R.A. = 23h21m24.6s
Dec. =	-04°17'25.4"	Dec. = -05°18'27.6"
S.D. =	00°16'06.6"	S.D. = 00°14'55.7"
H.P. =	00°00'08.9"	H.P. = 00°54'47.2"

External/Internal
Contacts of Penumbra

P1 = 10:30:40.9 UT1
P4 = 14:34:00.2 UT1

Circumstances at Greatest Eclipse

Lat. = 72°02.6'S	Sun Alt. = 0.0°
Long. = 097°52.1'E	Sun Azm. = 256.0°

ΔT = 87.8 s

www.EclipseWise.com

©2020 F. Espenak

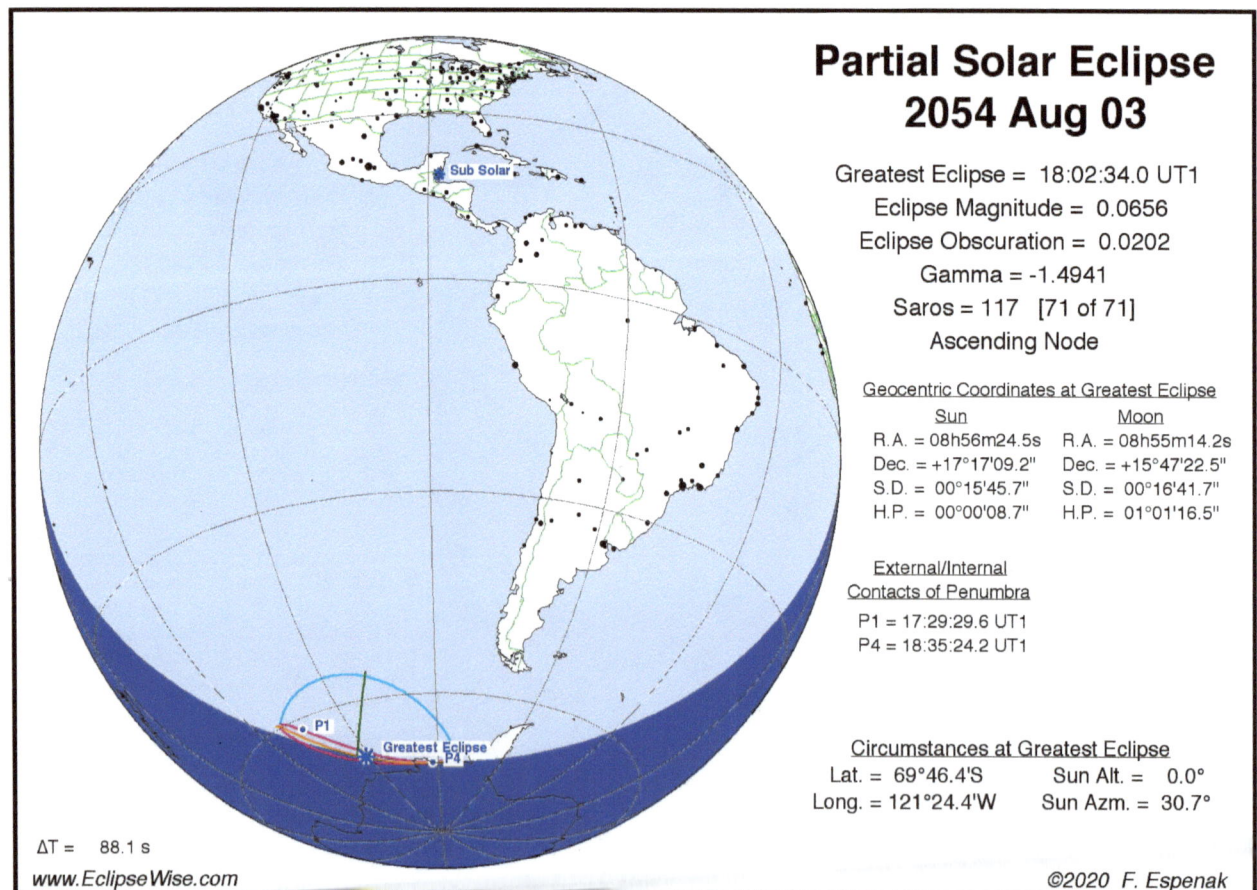

Partial Solar Eclipse
2054 Aug 03

Greatest Eclipse = 18:02:34.0 UT1

Eclipse Magnitude = 0.0656

Eclipse Obscuration = 0.0202

Gamma = -1.4941

Saros = 117   [71 of 71]

Ascending Node

Geocentric Coordinates at Greatest Eclipse

	Sun	Moon
R.A. =	08h56m24.5s	R.A. = 08h55m14.2s
Dec. =	+17°17'09.2"	Dec. = +15°47'22.5"
S.D. =	00°15'45.7"	S.D. = 00°16'41.7"
H.P. =	00°00'08.7"	H.P. = 01°01'16.5"

External/Internal
Contacts of Penumbra

P1 = 17:29:29.6 UT1
P4 = 18:35:24.2 UT1

Circumstances at Greatest Eclipse

Lat. = 69°46.4'S	Sun Alt. = 0.0°
Long. = 121°24.4'W	Sun Azm. = 30.7°

ΔT = 88.1 s

www.EclipseWise.com

©2020 F. Espenak

Partial Solar Eclipse
2054 Sep 02

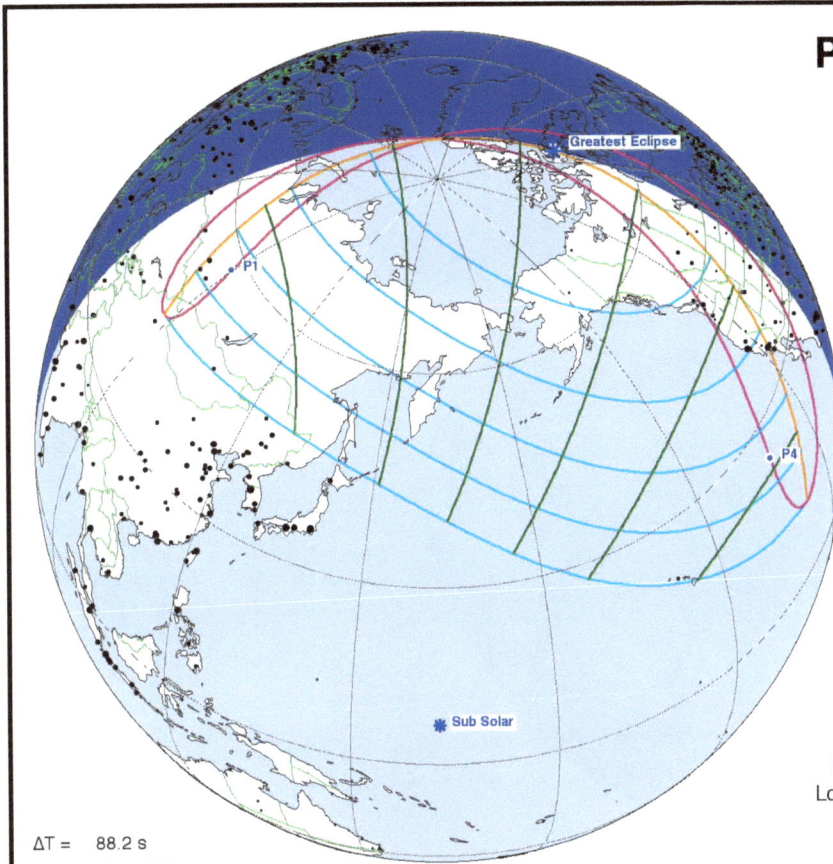

Greatest Eclipse = 01:08:05.5 UT1

Eclipse Magnitude = 0.9793

Eclipse Obscuration = 0.9828

Gamma = 1.0215

Saros = 155 [8 of 71]

Ascending Node

Geocentric Coordinates at Greatest Eclipse

	Sun	Moon
R.A. =	10h45m28.2s	10h46m40.4s
Dec. =	+07°52'58.6"	+08°52'49.8"
S.D. =	00°15'50.9"	00°16'42.0"
H.P. =	00°00'08.7"	01°01'17.5"

External/Internal Contacts of Penumbra

P1 = 23:10:53.0 UT1

P4 = 03:05:33.9 UT1

Circumstances at Greatest Eclipse

Lat. = 71°40.9'N	Sun Alt. = 0.0°
Long. = 082°24.5'W	Sun Azm. = 295.9°

ΔT =   88.2 s

www.EclipseWise.com

©2020 F. Espenak

Partial Solar Eclipse
2055 Jan 27

Greatest Eclipse = 17:52:36.8 UT1

Eclipse Magnitude = 0.6932

Eclipse Obscuration = 0.5966

Gamma = 1.1550

Saros = 122 [60 of 70]

Decending Node

Geocentric Coordinates at Greatest Eclipse

	Sun	Moon
R.A. =	20h40m41.0s	20h39m58.6s
Dec. =	-18°19'18.9"	-17°17'11.5"
S.D. =	00°16'14.5"	00°14'53.3"
H.P. =	00°00'08.9"	00°54'38.4"

External/Internal Contacts of Penumbra

P1 = 15:47:39.8 UT1

P4 = 19:57:27.6 UT1

Circumstances at Greatest Eclipse

Lat. = 69°32.9'N	Sun Alt. = 0.0°
Long. = 112°18.7'W	Sun Azm. = 154.1°

ΔT =   88.4 s

www.EclipseWise.com

©2020 F. Espenak

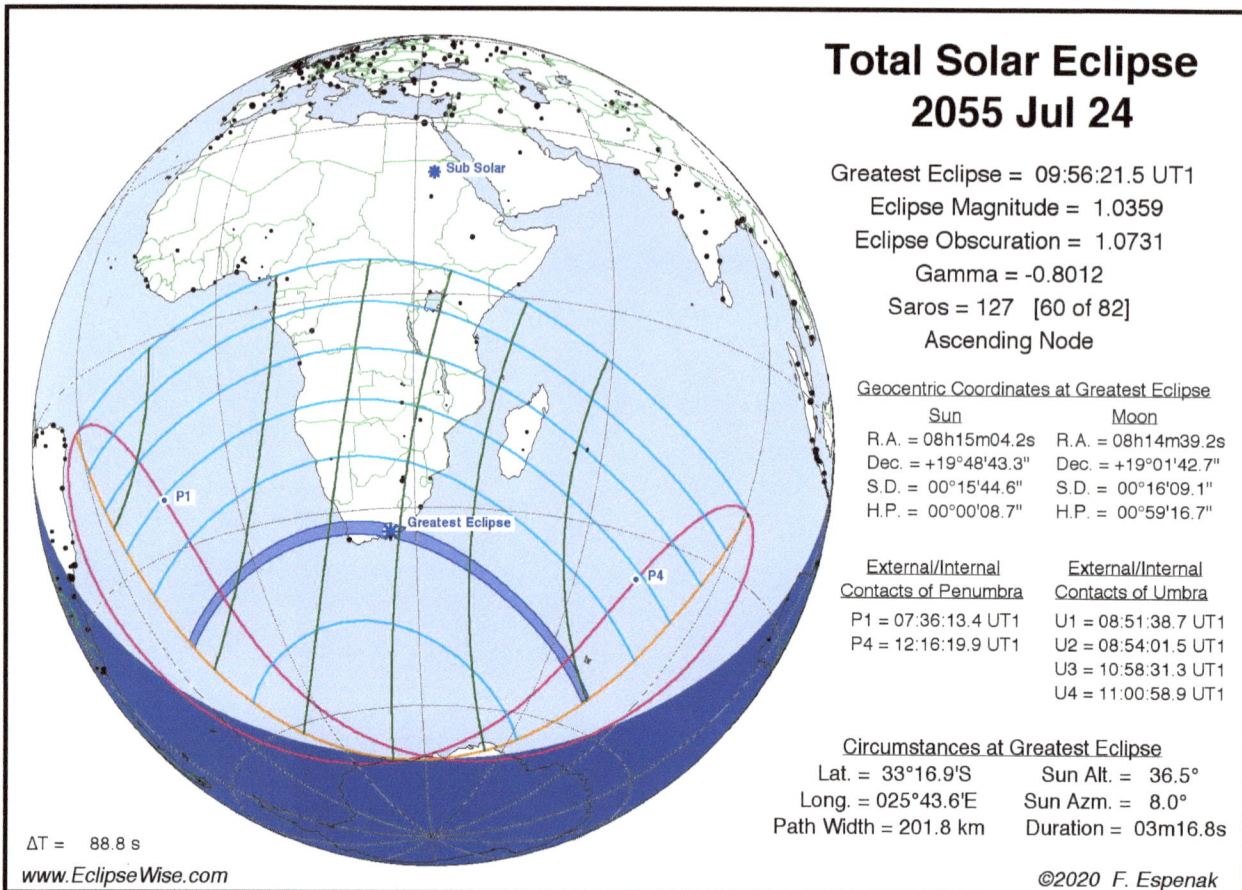

Total Solar Eclipse
2055 Jul 24

Greatest Eclipse = 09:56:21.5 UT1
Eclipse Magnitude = 1.0359
Eclipse Obscuration = 1.0731
Gamma = -0.8012
Saros = 127 [60 of 82]
Ascending Node

Geocentric Coordinates at Greatest Eclipse

Sun	Moon
R.A. = 08h15m04.2s	R.A. = 08h14m39.2s
Dec. = +19°48'43.3"	Dec. = +19°01'42.7"
S.D. = 00°15'44.6"	S.D. = 00°16'09.1"
H.P. = 00°00'08.7"	H.P. = 00°59'16.7"

External/Internal Contacts of Penumbra	External/Internal Contacts of Umbra
P1 = 07:36:13.4 UT1	U1 = 08:51:38.7 UT1
P4 = 12:16:19.9 UT1	U2 = 08:54:01.5 UT1
	U3 = 10:58:31.3 UT1
	U4 = 11:00:58.9 UT1

Circumstances at Greatest Eclipse

Lat. = 33°16.9'S	Sun Alt. = 36.5°
Long. = 025°43.6'E	Sun Azm. = 8.0°
Path Width = 201.8 km	Duration = 03m16.8s

ΔT = 88.8 s

www.EclipseWise.com

©2020 F. Espenak

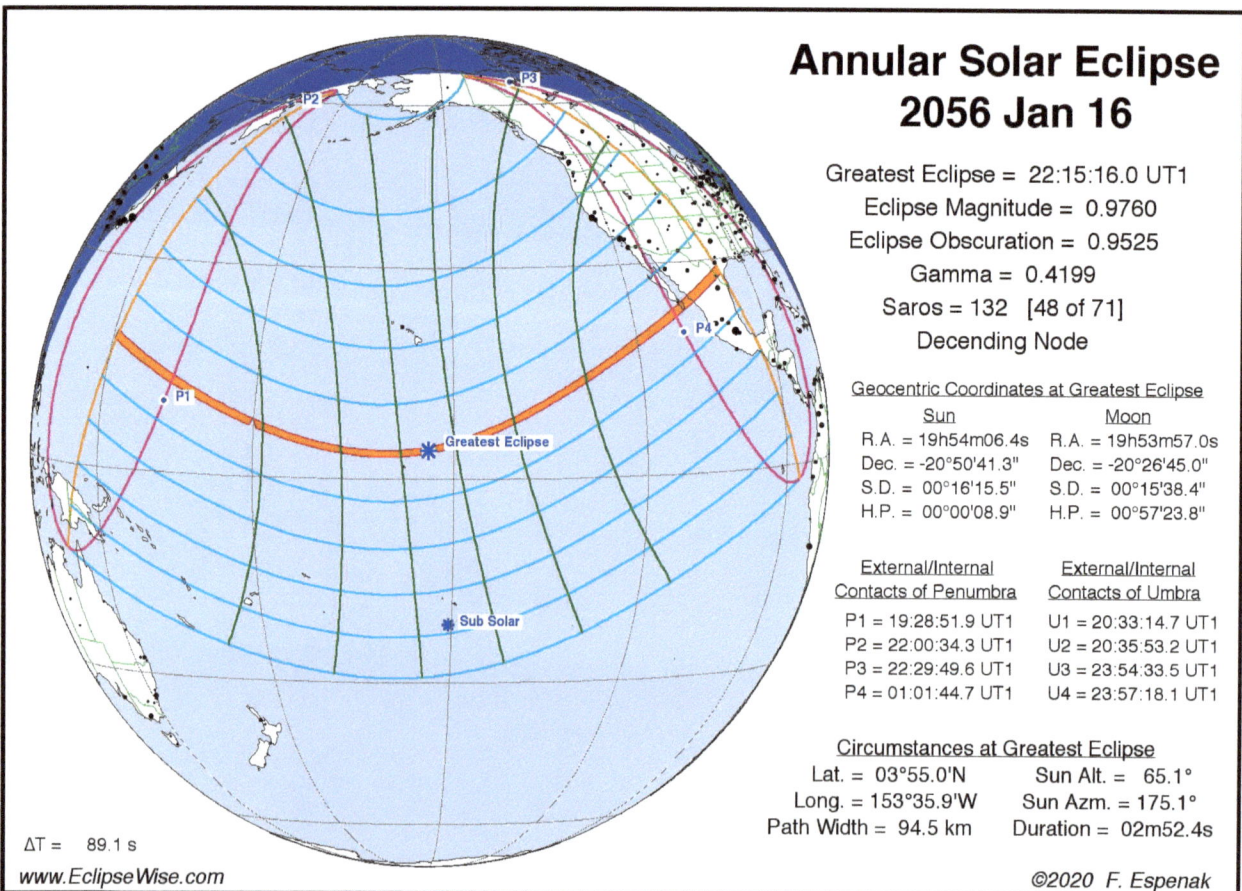

Annular Solar Eclipse
2056 Jan 16

Greatest Eclipse = 22:15:16.0 UT1
Eclipse Magnitude = 0.9760
Eclipse Obscuration = 0.9525
Gamma = 0.4199
Saros = 132 [48 of 71]
Decending Node

Geocentric Coordinates at Greatest Eclipse

Sun	Moon
R.A. = 19h54m06.4s	R.A. = 19h53m57.0s
Dec. = -20°50'41.3"	Dec. = -20°26'45.0"
S.D. = 00°16'15.5"	S.D. = 00°15'38.4"
H.P. = 00°00'08.9"	H.P. = 00°57'23.8"

External/Internal Contacts of Penumbra	External/Internal Contacts of Umbra
P1 = 19:28:51.9 UT1	U1 = 20:33:14.7 UT1
P2 = 22:00:34.3 UT1	U2 = 20:35:53.2 UT1
P3 = 22:29:49.6 UT1	U3 = 23:54:33.5 UT1
P4 = 01:01:44.7 UT1	U4 = 23:57:18.1 UT1

Circumstances at Greatest Eclipse

Lat. = 03°55.0'N	Sun Alt. = 65.1°
Long. = 153°35.9'W	Sun Azm. = 175.1°
Path Width = 94.5 km	Duration = 02m52.4s

ΔT = 89.1 s

www.EclipseWise.com

©2020 F. Espenak

Annular Solar Eclipse
2056 Jul 12

Greatest Eclipse = 20:20:29.9 UT1

Eclipse Magnitude = 0.9878

Eclipse Obscuration = 0.9757

Gamma = -0.0426

Saros = 137   [38 of 70]

Ascending Node

Geocentric Coordinates at Greatest Eclipse

Sun	Moon
R.A. = 07h31m42.7s	R.A. = 07h31m42.1s
Dec. = +21°45'32.5"	Dec. = +21°43'09.5"
S.D. = 00°15'44.0"	S.D. = 00°15'17.9"
H.P. = 00°00'08.7"	H.P. = 00°56'08.9"

External/Internal Contacts of Penumbra	External/Internal Contacts of Umbra
P1 = 17:24:26.3 UT1	U1 = 18:26:22.9 UT1
P2 = 19:30:13.7 UT1	U2 = 18:28:13.4 UT1
P3 = 21:10:46.8 UT1	U3 = 22:12:48.2 UT1
P4 = 23:16:27.1 UT1	U4 = 22:14:33.1 UT1

Circumstances at Greatest Eclipse

Lat. = 19°26.4'N	Sun Alt. = 87.7°
Long. = 123°48.1'W	Sun Azm. = 3.3°
Path Width = 43.2 km	Duration = 01m25.9s

ΔT =   89.4 s

www.EclipseWise.com

©2020 F. Espenak

Total Solar Eclipse
2057 Jan 05

Greatest Eclipse = 09:46:22.4 UT1

Eclipse Magnitude = 1.0287

Eclipse Obscuration = 1.0583

Gamma = -0.2837

Saros = 142   [25 of 72]

Decending Node

Geocentric Coordinates at Greatest Eclipse

Sun	Moon
R.A. = 19h07m25.3s	R.A. = 19h07m26.6s
Dec. = -22°31'37.8"	Dec. = -22°48'43.6"
S.D. = 00°16'15.9"	S.D. = 00°16'27.8"
H.P. = 00°00'08.9"	H.P. = 01°00'25.4"

External/Internal Contacts of Penumbra	External/Internal Contacts of Umbra
P1 = 07:07:22.9 UT1	U1 = 08:05:30.1 UT1
P2 = 09:08:57.7 UT1	U2 = 08:06:13.0 UT1
P3 = 10:23:47.0 UT1	U3 = 11:26:34.2 UT1
P4 = 12:25:25.9 UT1	U4 = 11:27:13.6 UT1

Circumstances at Greatest Eclipse

Lat. = 39°13.0'S	Sun Alt. = 73.3°
Long. = 035°07.1'E	Sun Azm. = 358.9°
Path Width = 101.6 km	Duration = 02m28.7s

ΔT =   89.8 s

www.EclipseWise.com

©2020 F. Espenak

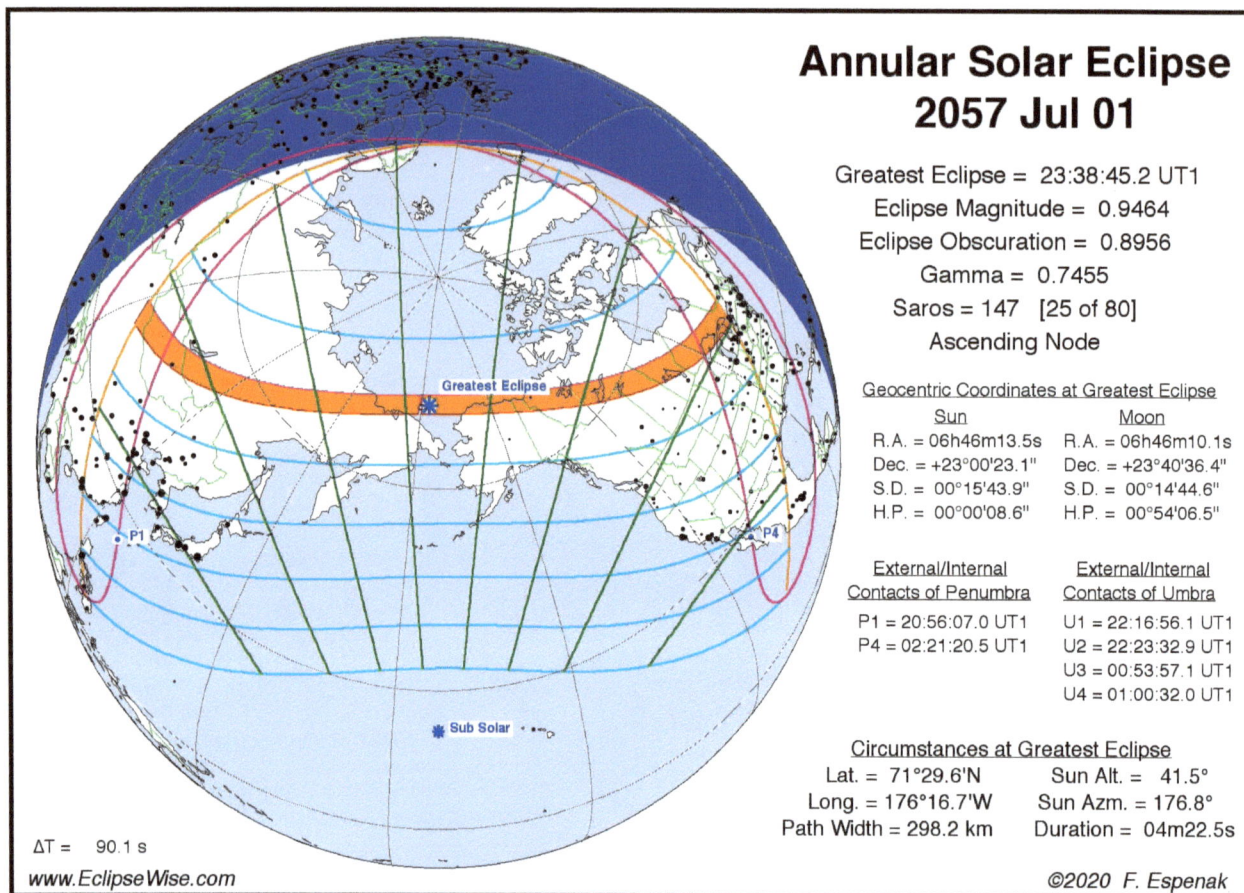

Annular Solar Eclipse
2057 Jul 01

Greatest Eclipse = 23:38:45.2 UT1

Eclipse Magnitude = 0.9464

Eclipse Obscuration = 0.8956

Gamma = 0.7455

Saros = 147 [25 of 80]

Ascending Node

Geocentric Coordinates at Greatest Eclipse

Sun	Moon
R.A. = 06h46m13.5s	R.A. = 06h46m10.1s
Dec. = +23°00'23.1"	Dec. = +23°40'36.4"
S.D. = 00°15'43.9"	S.D. = 00°14'44.6"
H.P. = 00°00'08.6"	H.P. = 00°54'06.5"

External/Internal Contacts of Penumbra	External/Internal Contacts of Umbra
P1 = 20:56:07.0 UT1	U1 = 22:16:56.1 UT1
P4 = 02:21:20.5 UT1	U2 = 22:23:32.9 UT1
	U3 = 00:53:57.1 UT1
	U4 = 01:00:32.0 UT1

Circumstances at Greatest Eclipse

Lat. = 71°29.6'N	Sun Alt. = 41.5°
Long. = 176°16.7'W	Sun Azm. = 176.8°
Path Width = 298.2 km	Duration = 04m22.5s

ΔT = 90.1 s

www.EclipseWise.com

©2020 F. Espenak

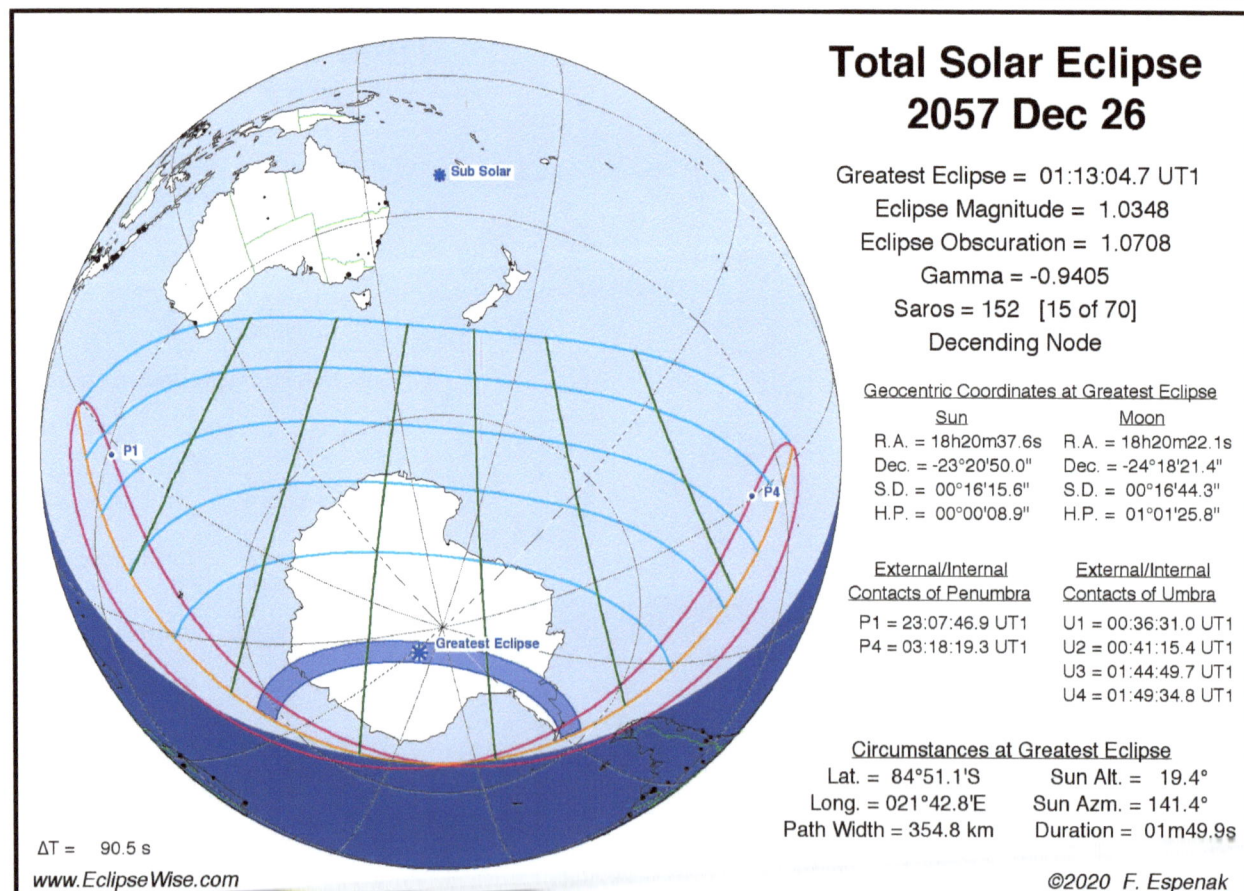

Total Solar Eclipse
2057 Dec 26

Greatest Eclipse = 01:13:04.7 UT1

Eclipse Magnitude = 1.0348

Eclipse Obscuration = 1.0708

Gamma = -0.9405

Saros = 152 [15 of 70]

Decending Node

Geocentric Coordinates at Greatest Eclipse

Sun	Moon
R.A. = 18h20m37.6s	R.A. = 18h20m22.1s
Dec. = -23°20'50.0"	Dec. = -24°18'21.4"
S.D. = 00°16'15.6"	S.D. = 00°16'44.3"
H.P. = 00°00'08.9"	H.P. = 01°01'25.8"

External/Internal Contacts of Penumbra	External/Internal Contacts of Umbra
P1 = 23:07:46.9 UT1	U1 = 00:36:31.0 UT1
P4 = 03:18:19.3 UT1	U2 = 00:41:15.4 UT1
	U3 = 01:44:49.7 UT1
	U4 = 01:49:34.8 UT1

Circumstances at Greatest Eclipse

Lat. = 84°51.1'S	Sun Alt. = 19.4°
Long. = 021°42.8'E	Sun Azm. = 141.4°
Path Width = 354.8 km	Duration = 01m49.9s

ΔT = 90.5 s

www.EclipseWise.com

©2020 F. Espenak

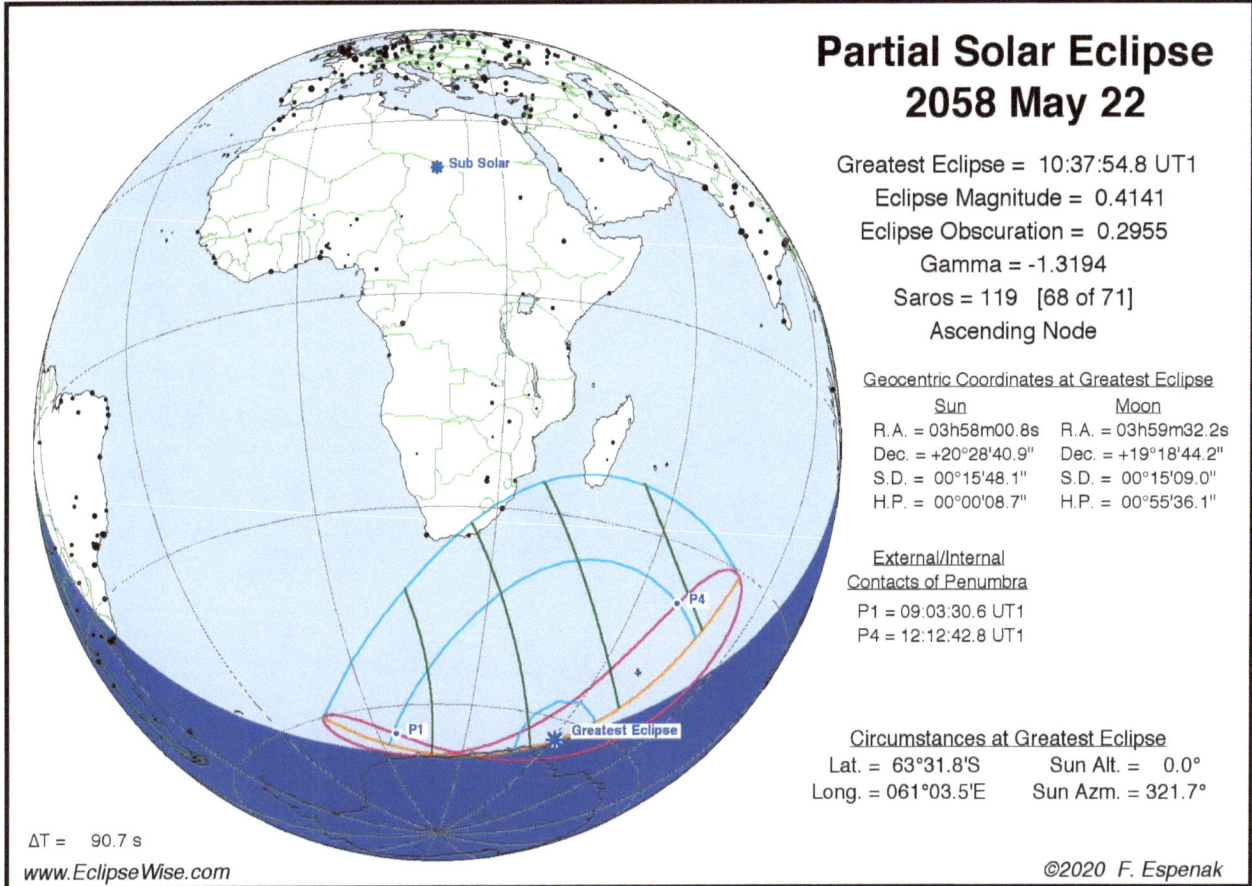

Partial Solar Eclipse
2058 May 22

Greatest Eclipse = 10:37:54.8 UT1

Eclipse Magnitude = 0.4141

Eclipse Obscuration = 0.2955

Gamma = -1.3194

Saros = 119 [68 of 71]

Ascending Node

Geocentric Coordinates at Greatest Eclipse

	Sun	Moon
R.A. =	03h58m00.8s	03h59m32.2s
Dec. =	+20°28'40.9"	+19°18'44.2"
S.D. =	00°15'48.1"	00°15'09.0"
H.P. =	00°00'08.7"	00°55'36.1"

External/Internal Contacts of Penumbra

P1 = 09:03:30.6 UT1
P4 = 12:12:42.8 UT1

Circumstances at Greatest Eclipse

Lat. = 63°31.8'S	Sun Alt. = 0.0°
Long. = 061°03.5'E	Sun Azm. = 321.7°

ΔT =   90.7 s

www.EclipseWise.com

©2020 F. Espenak

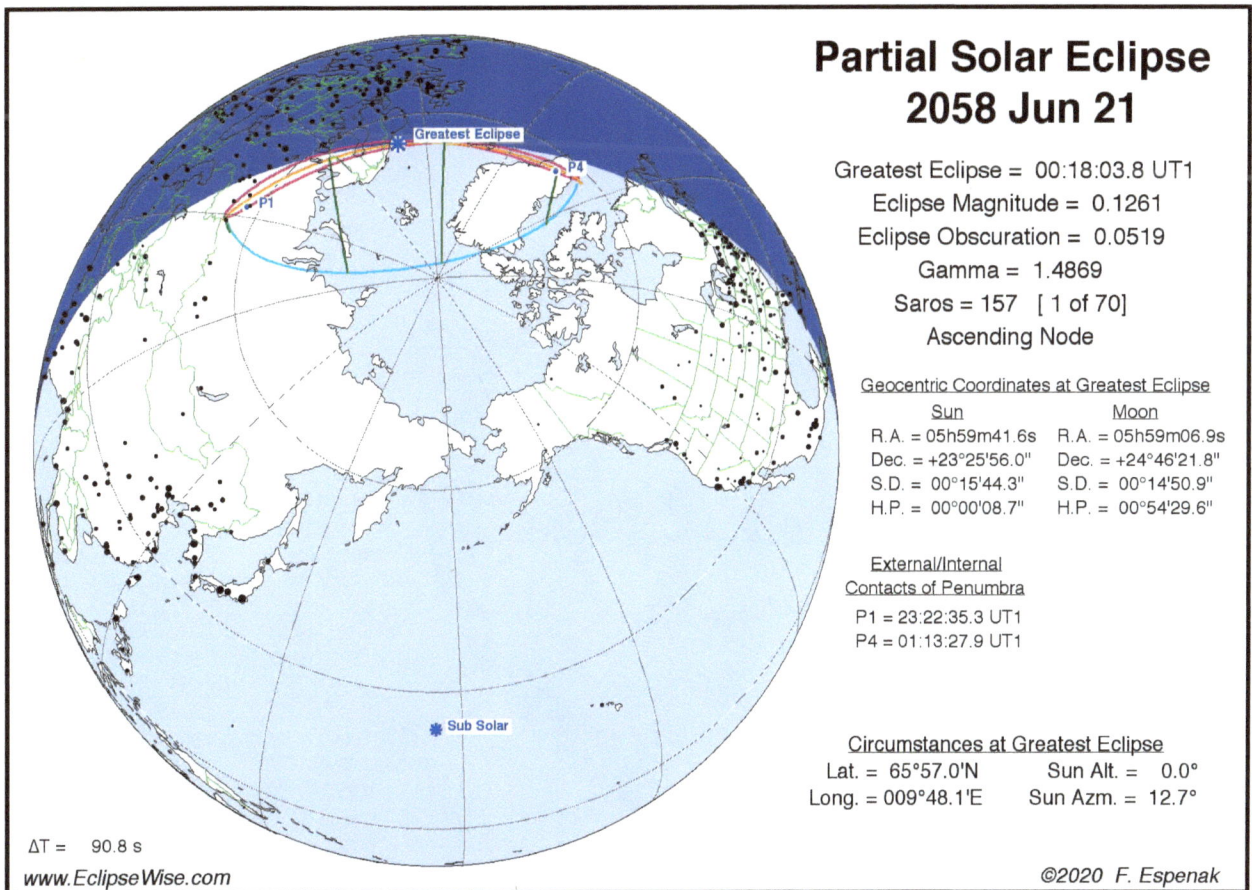

Partial Solar Eclipse
2058 Jun 21

Greatest Eclipse = 00:18:03.8 UT1

Eclipse Magnitude = 0.1261

Eclipse Obscuration = 0.0519

Gamma = 1.4869

Saros = 157   [1 of 70]

Ascending Node

Geocentric Coordinates at Greatest Eclipse

	Sun	Moon
R.A. =	05h59m41.6s	05h59m06.9s
Dec. =	+23°25'56.0"	+24°46'21.8"
S.D. =	00°15'44.3"	00°14'50.9"
H.P. =	00°00'08.7"	00°54'29.6"

External/Internal Contacts of Penumbra

P1 = 23:22:35.3 UT1
P4 = 01:13:27.9 UT1

Circumstances at Greatest Eclipse

Lat. = 65°57.0'N	Sun Alt. = 0.0°
Long. = 009°48.1'E	Sun Azm. = 12.7°

ΔT =   90.8 s

www.EclipseWise.com

©2020 F. Espenak

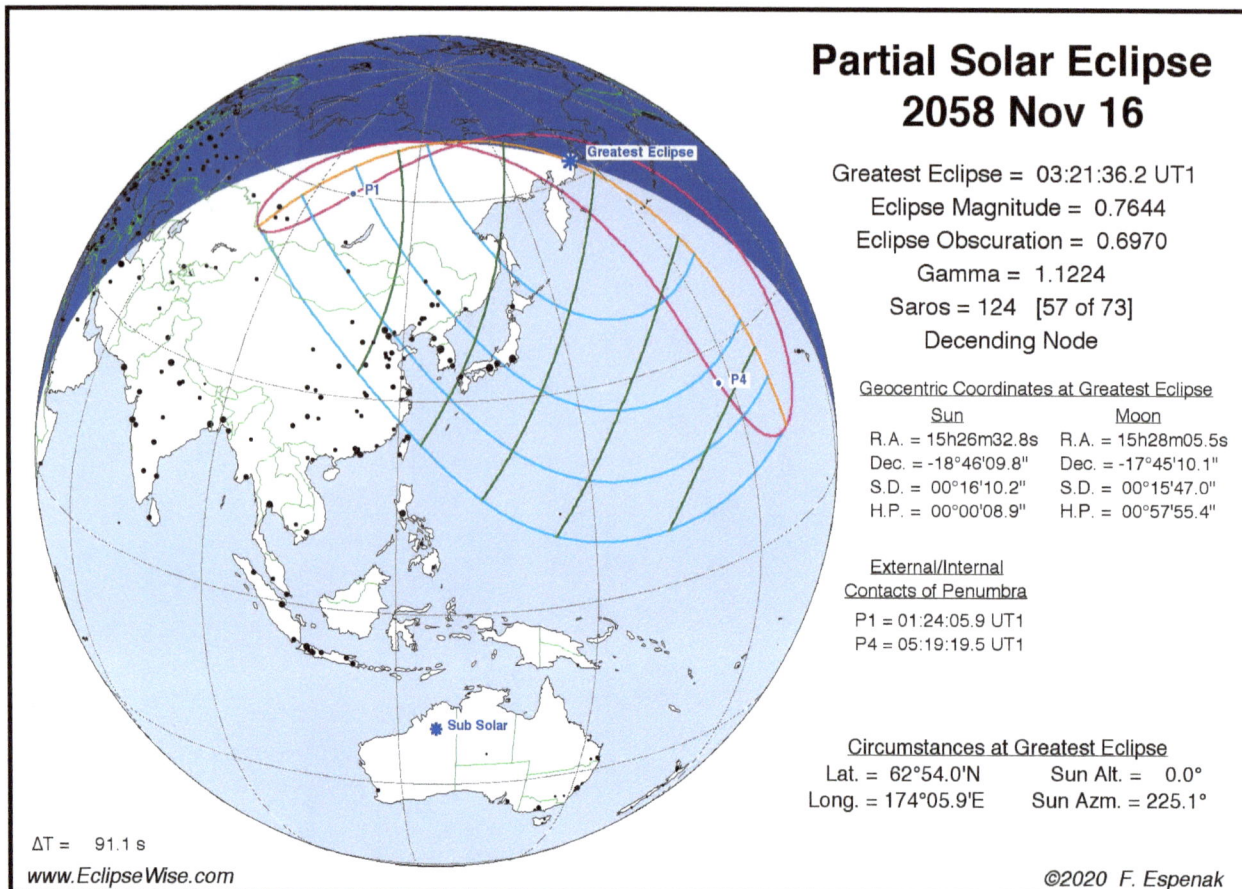

Partial Solar Eclipse
2058 Nov 16

Greatest Eclipse = 03:21:36.2 UT1

Eclipse Magnitude = 0.7644

Eclipse Obscuration = 0.6970

Gamma = 1.1224

Saros = 124 [57 of 73]

Decending Node

Geocentric Coordinates at Greatest Eclipse

Sun	Moon
R.A. = 15h26m32.8s	R.A. = 15h28m05.5s
Dec. = -18°46'09.8"	Dec. = -17°45'10.1"
S.D. = 00°16'10.2"	S.D. = 00°15'47.0"
H.P. = 00°00'08.9"	H.P. = 00°57'55.4"

External/Internal Contacts of Penumbra

P1 = 01:24:05.9 UT1
P4 = 05:19:19.5 UT1

Circumstances at Greatest Eclipse

Lat. = 62°54.0'N	Sun Alt. = 0.0°
Long. = 174°05.9'E	Sun Azm. = 225.1°

ΔT = 91.1 s

www.EclipseWise.com

©2020 F. Espenak

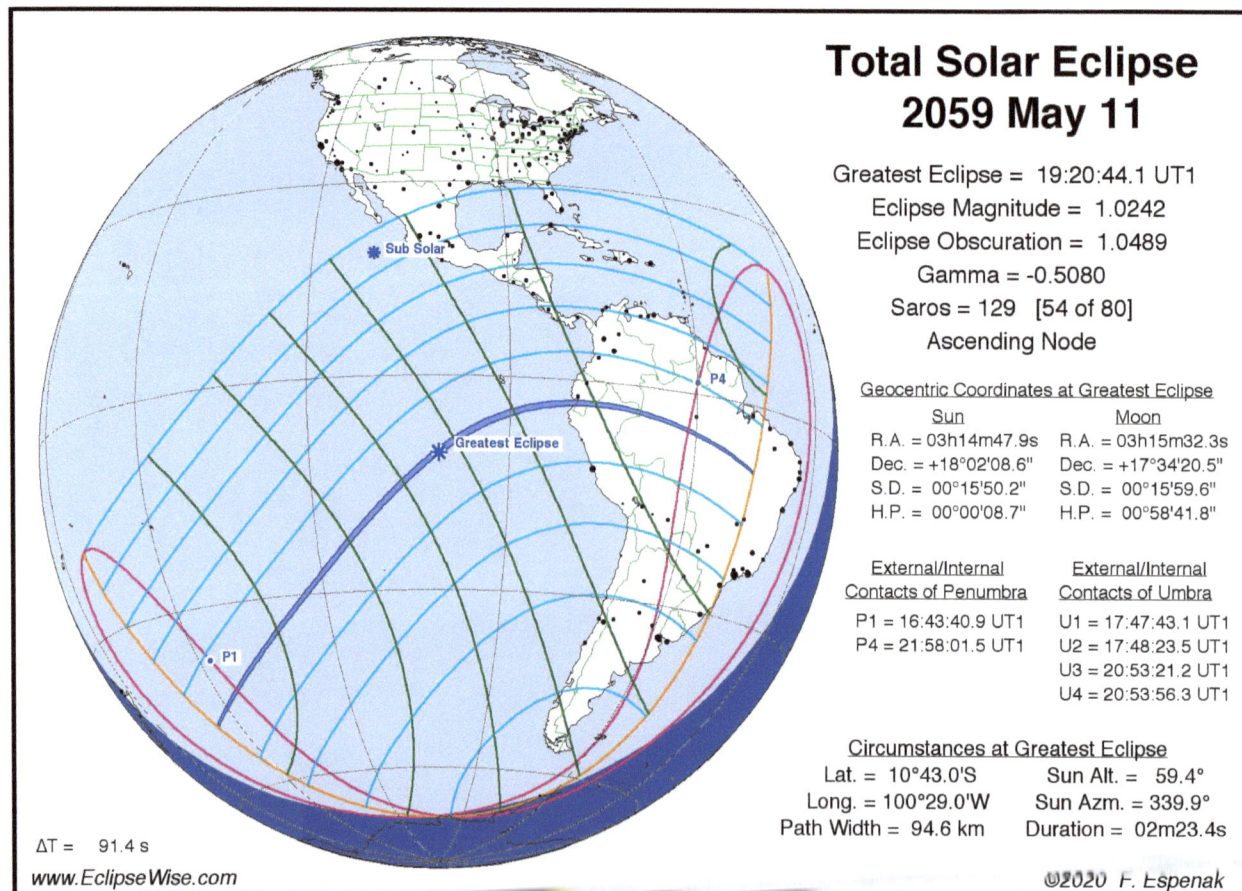

Total Solar Eclipse
2059 May 11

Greatest Eclipse = 19:20:44.1 UT1

Eclipse Magnitude = 1.0242

Eclipse Obscuration = 1.0489

Gamma = -0.5080

Saros = 129 [54 of 80]

Ascending Node

Geocentric Coordinates at Greatest Eclipse

Sun	Moon
R.A. = 03h14m47.9s	R.A. = 03h15m32.3s
Dec. = +18°02'08.6"	Dec. = +17°34'20.5"
S.D. = 00°15'50.2"	S.D. = 00°15'59.6"
H.P. = 00°00'08.7"	H.P. = 00°58'41.8"

External/Internal Contacts of Penumbra / Contacts of Umbra

External/Internal Contacts of Penumbra	External/Internal Contacts of Umbra
P1 = 16:43:40.9 UT1	U1 = 17:47:43.1 UT1
P4 = 21:58:01.5 UT1	U2 = 17:48:23.5 UT1
	U3 = 20:53:21.2 UT1
	U4 = 20:53:56.3 UT1

Circumstances at Greatest Eclipse

Lat. = 10°43.0'S	Sun Alt. = 59.4°
Long. = 100°29.0'W	Sun Azm. = 339.9°
Path Width = 94.6 km	Duration = 02m23.4s

ΔT = 91.4 s

www.EclipseWise.com

©2020 F. Espenak

Annular Solar Eclipse
2059 Nov 05

Greatest Eclipse = 09:16:42.8 UT1

Eclipse Magnitude = 0.9417

Eclipse Obscuration = 0.8867

Gamma = 0.4454

Saros = 134 [46 of 71]

Decending Node

Geocentric Coordinates at Greatest Eclipse

Sun	Moon
R.A. = 14h42m02.6s	R.A. = 14h42m42.7s
Dec. = -15°43'28.3"	Dec. = -15°21'02.7"
S.D. = 00°16'07.6"	S.D. = 00°14'58.8"
H.P. = 00°00'08.9"	H.P. = 00°54'58.7"

External/Internal Contacts of Penumbra	External/Internal Contacts of Umbra
P1 = 06:21:44.5 UT1	U1 = 07:29:52.3 UT1
P4 = 12:11:45.2 UT1	U2 = 07:35:24.6 UT1
	U3 = 10:58:17.2 UT1
	U4 = 11:03:44.4 UT1

Circumstances at Greatest Eclipse

Lat. = 08°44.6'N	Sun Alt. = 63.5°
Long. = 046°59.7'E	Sun Azm. = 202.7°
Path Width = 238.3 km	Duration = 06m59.6s

ΔT = 91.8 s

www.EclipseWise.com

©2020 F. Espenak

Total Solar Eclipse
2060 Apr 30

Greatest Eclipse = 10:08:27.7 UT1

Eclipse Magnitude = 1.0660

Eclipse Obscuration = 1.1364

Gamma = 0.2422

Saros = 139 [32 of 71]

Ascending Node

Geocentric Coordinates at Greatest Eclipse

Sun	Moon
R.A. = 02h33m38.4s	R.A. = 02h33m13.6s
Dec. = +15°04'16.7"	Dec. = +15°17'46.8"
S.D. = 00°15'52.6"	S.D. = 00°16'38.8"
H.P. = 00°00'08.7"	H.P. = 01°01'05.8"

External/Internal Contacts of Penumbra	External/Internal Contacts of Umbra
P1 = 07:32:05.2 UT1	U1 = 08:26:56.0 UT1
P2 = 09:27:18.4 UT1	U2 = 08:29:36.1 UT1
P3 = 10:49:21.7 UT1	U3 = 11:47:13.4 UT1
P4 = 12:44:47.6 UT1	U4 = 11:49:51.8 UT1

Circumstances at Greatest Eclipse

Lat. = 27°57.4'N	Sun Alt. = 75.8°
Long. = 020°49.4'E	Sun Azm. = 154.1°
Path Width = 222.0 km	Duration = 05m15.2s

ΔT = 92.1 s

www.EclipseWise.com

©2020 F. Espenak

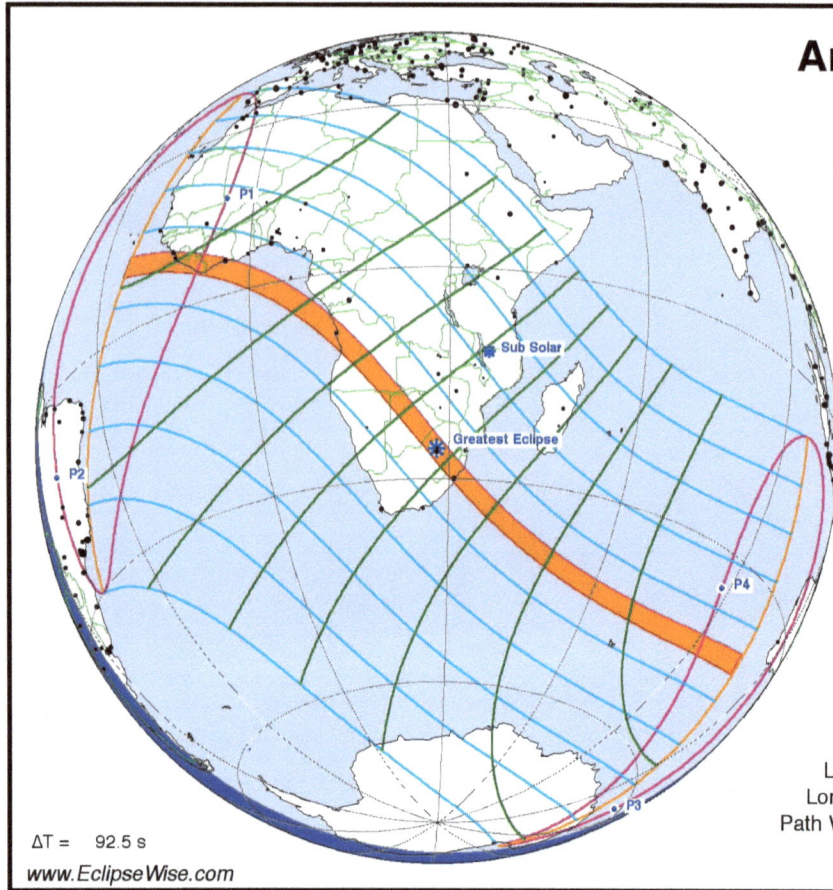

Annular Solar Eclipse
2060 Oct 24

Greatest Eclipse = 09:22:37.9 UT1

Eclipse Magnitude = 0.9277

Eclipse Obscuration = 0.8606

Gamma = -0.2625

Saros = 144 [19 of 70]

Decending Node

Geocentric Coordinates at Greatest Eclipse

	Sun	Moon
R.A. =	13h58m17.5s	13h57m52.2s
Dec. =	-12°04'28.2"	-12°17'09.7"
S.D. =	00°16'04.8"	00°14'42.1"
H.P. =	00°00'08.8"	00°53'57.3"

External/Internal Contacts of Penumbra	External/Internal Contacts of Umbra
P1 = 06:18:08.0 UT1	U1 = 07:24:35.7 UT1
P2 = 08:42:25.1 UT1	U2 = 07:30:56.7 UT1
P3 = 10:02:28.2 UT1	U3 = 11:14:08.9 UT1
P4 = 12:27:02.8 UT1	U4 = 11:20:31.4 UT1

Circumstances at Greatest Eclipse

Lat. = 25°47.2'S	Sun Alt. = 74.6°
Long. = 028°01.5'E	Sun Azm. = 28.1°
Path Width = 280.5 km	Duration = 08m05.8s

ΔT = 92.5 s

Key to Catalog of Solar Eclipses

The following catalog lists all solar eclipses for a 25-year period.

A brief description of each parameter in the catalog appears below.

Date — Gregorian date of Greatest Eclipse

Greatest Eclipse — Universal Time of Greatest Eclipse

Saros — Saros Series Number of the eclipse

Type — Solar Eclipse Type

      P = Partial Solar Eclipse
      A = Annular Solar Eclipse
      T = Total Solar Eclipse
      H = Hybrid Solar Eclipse

      + = Non-central eclipse of with no northern limit
      – = Non-central eclipse of with no southern limit

      b = Saros series begins (first eclipse in a Saros series)
      e = Saros series ends (last eclipse in a Saros series)

Gamma — minimum distance from the axis of the lunar shadow to the center of Earth

Mag — Eclipse Magnitude; fraction of the Sun's diameter obscured by the Moon

Lat & Long — latitude and longitude where the Sun appears in the zenith at greatest eclipse

Alt — altitude of the Sun at greatest eclipse

Duration — Central Line Duration (minutes. seconds) at greatest eclipse

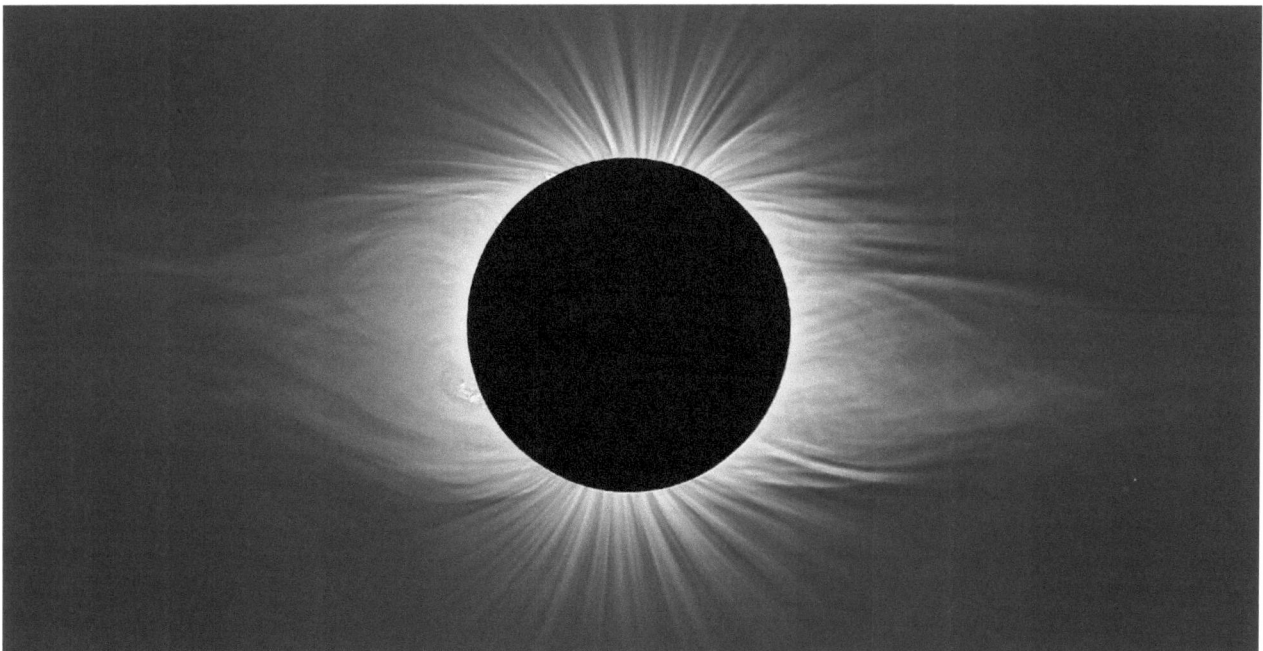

Photo 1–8 The solar corona is revealed during totality. Total Solar Eclipse of 2018 Jul 03. ©2018 F. Espenak

Catalog of Solar Eclipses: 2051 to 2085

Date	Greatest Eclipse	Saros	Type	Gamma	Mag	Lat	Long	Alt	Duration
2051 Apr 11	2:09:13	120	P	1.0169	0.9849	71.6N	32.1E	0	
2051 Oct 04	21:00:48	125	P	-1.2094	0.6024	72.0S	117.7E	0	
2052 Mar 30	18:30:26	130	T	0.3239	1.0466	22.4N	102.6W	71	04m08s
2052 Sep 22	23:37:43	135	A	-0.4480	0.9734	25.7S	174.9E	63	02m51s
2053 Mar 20	7:06:52	140	A	-0.4089	0.9919	23.0S	82.9E	66	00m50s
2053 Sep 12	9:32:41	145	T	0.3140	1.0328	21.5N	41.7E	72	03m04s
2054 Mar 09	12:32:13	150	P	-1.1711	0.6678	72.0S	97.9E	0	
2054 Aug 03	18:02:34	117	Pe	-1.4941	0.0656	69.8S	121.4W	0	
2054 Sep 02	1:08:05	155	P	1.0215	0.9793	71.7N	82.4W	0	
2055 Jan 27	17:52:37	122	P	1.1550	0.6932	69.5N	112.3W	0	
2055 Jul 24	9:56:21	127	T	-0.8012	1.0359	33.3S	25.7E	37	03m17s
2056 Jan 16	22:15:16	132	A	0.4199	0.9760	3.9N	153.6W	65	02m52s
2056 Jul 12	20:20:30	137	A	-0.0426	0.9878	19.4N	123.8W	88	01m26s
2057 Jan 05	9:46:22	142	T	-0.2837	1.0287	39.2S	35.1E	73	02m29s
2057 Jul 01	23:38:45	147	A	0.7455	0.9464	71.5N	176.3W	41	04m22s
2057 Dec 26	1:13:05	152	T	-0.9405	1.0348	84.9S	21.7E	19	01m50s
2058 May 22	10:37:55	119	P	-1.3194	0.4141	63.5S	61.1E	0	
2058 Jun 21	0:18:04	157	Pb	1.4869	0.1261	65.9N	9.8E	0	
2058 Nov 16	3:21:36	124	P	1.1224	0.7644	62.9N	174.1E	0	
2059 May 11	19:20:44	129	T	-0.5080	1.0242	10.7S	100.5W	59	02m23s
2059 Nov 05	9:16:43	134	A	0.4454	0.9417	8.7N	47.0E	63	07m00s
2060 Apr 30	10:08:28	139	T	0.2422	1.0660	28.0N	20.8E	76	05m15s
2060 Oct 24	9:22:38	144	A	-0.2625	0.9277	25.8S	28.0E	75	08m06s
2061 Apr 20	2:55:16	149	T	0.9578	1.0476	64.5N	59.1E	16	02m37s
2061 Oct 13	10:30:36	154	A	-0.9639	0.9469	62.1S	54.5W	15	03m41s
2062 Mar 11	4:24:43	121	P	-1.0238	0.9331	61.0S	147.2W	0	
2062 Sep 03	8:52:54	126	P	1.0192	0.9749	61.3N	150.2E	0	
2063 Feb 28	7:41:56	131	A	-0.3360	0.9293	25.2S	77.6E	70	07m41s
2063 Aug 24	1:20:36	136	T	0.2771	1.0750	25.6N	168.3E	74	05m49s
2064 Feb 17	6:58:48	141	A	0.3596	0.9262	7.0N	69.6E	69	08m56s
2064 Aug 12	17:44:31	146	T	-0.4652	1.0495	10.9S	96.1W	62	04m28s
2065 Feb 05	9:50:50	151	P	1.0336	0.9123	62.2N	22.0W	0	
2065 Jul 03	17:32:17	118	P	1.4619	0.1639	64.8N	71.7E	0	
2065 Aug 02	5:32:41	156	P	-1.2758	0.4903	62.7S	46.4E	0	
2065 Dec 27	8:38:19	123	P	-1.0688	0.8769	65.4S	149.3W	0	
2066 Jun 22	19:24:11	128	A	0.7330	0.9435	70.1N	96.5W	43	04m40s
2066 Dec 17	0:22:03	133	T	-0.4043	1.0416	47.4S	175.6E	66	03m14s
2067 Jun 11	20:40:49	138	A	-0.0387	0.9670	21.0N	130.3W	88	04m05s
2067 Dec 06	14:02:05	143	H	0.2845	1.0011	6.0S	32.5W	74	00m08s
2068 May 31	3:55:01	148	T	-0.7970	1.0110	31.0S	123.1E	37	01m06s
2068 Nov 24	21:30:51	153	P	1.0299	0.9109	68.5N	131.2W	0	
2069 Apr 21	10:09:30	120	P	1.0624	0.8992	71.0N	101.4W	0	
2069 May 20	17:51:39	158	Pb	-1.4852	0.0879	68.8S	70.1W	0	
2069 Oct 15	4:18:17	125	P	-1.2524	0.5298	71.6S	5.6W	0	
2070 Apr 11	2:34:30	130	T	0.3652	1.0472	29.1N	134.9E	68	04m04s
2070 Oct 04	7:07:17	135	A	-0.4950	0.9731	32.8S	60.3E	60	02m44s
2071 Mar 31	14:59:26	140	A	-0.3739	0.9919	16.7S	37.2W	68	00m52s
2071 Sep 23	17:18:47	145	T	0.2620	1.0333	14.2N	76.9W	75	03m11s
2072 Mar 19	20:08:50	150	P	-1.1405	0.7199	72.2S	30.5W	0	
2072 Sep 12	8:57:39	155	T	0.9655	1.0558	69.8N	101.8E	14	03m13s
2073 Feb 07	1:54:17	122	P	1.1651	0.6768	70.5N	114.7E	0	
2073 Aug 03	17:13:41	127	T	-0.8763	1.0294	43.2S	89.6W	28	02m29s
2074 Jan 27	6:42:33	132	A	0.4251	0.9798	6.6N	78.6E	65	02m21s
2074 Jul 24	3:08:49	137	A	-0.1242	0.9838	12.8N	133.6E	83	01m57s
2075 Jan 16	18:34:21	142	T	-0.2799	1.0311	37.2S	94.3W	74	02m42s
2075 Jul 13	6:04:00	117	A	0.8583	0.9467	63.1N	95.0E	49	04m45s

Photo 2–1 Phases of the total lunar eclipse of 2014 Apr 15 are captured in this multiple exposure sequence. ©2014 F. Espenak

Section 2: Lunar Eclipses

Introduction

The Moon orbits Earth once every 29.5306 days with respect to the Sun. Over the course of its orbit, the Moon's changing position relative to the Sun results in its familiar phases: New Moon > First Quarter > Full Moon > Last Quarter > New Moon. The New Moon phase is the only one not visible because the illuminated side of the Moon points away from Earth.

During Full Moon, the Moon appears opposite the Sun in the sky. It rises as the Sun sets and is visible throughout the night. The Full Moon sets in the morning just as the Sun rises. This geometry occurs when the Moon is 180° from the Sun as seen from Earth. It corresponds to the direction Earth casts its shadow into space.

The Moon's orbit is tilted about 5.1° to Earth's orbit around the Sun. As seen from Earth, the points where the two orbits appear to cross are called the nodes. When the Full Moon occurs near one of these nodes, the Moon can pass through some portion of Earth's shadow and a lunar eclipse occurs.

Earth's shadow has two cone-shaped components, one nested inside the other. The outer or penumbral shadow is a zone where the Sun's rays are partially blocked., The inner or umbral shadow is a region where direct rays from the Sun are completely blocked.

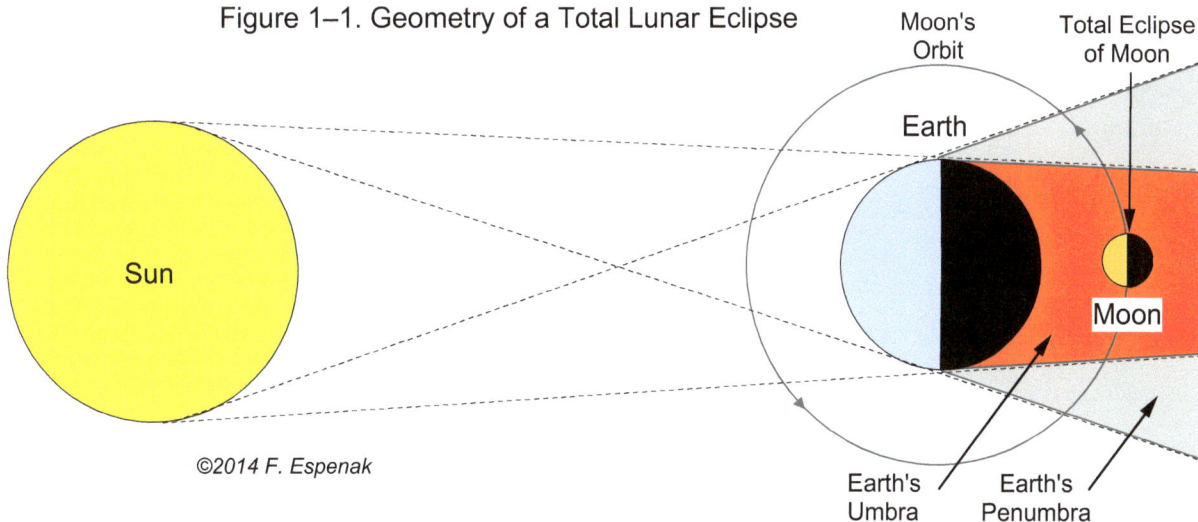

Figure 1–1. Geometry of a Total Lunar Eclipse

Figure 2–1 illustrates the geometry of a total lunar eclipse. A partial eclipse is visible if only part of the Moon enters Earth's umbral shadow. If the Moon passes through the penumbral shadow but misses the umbral shadow, then a penumbral eclipse occurs.

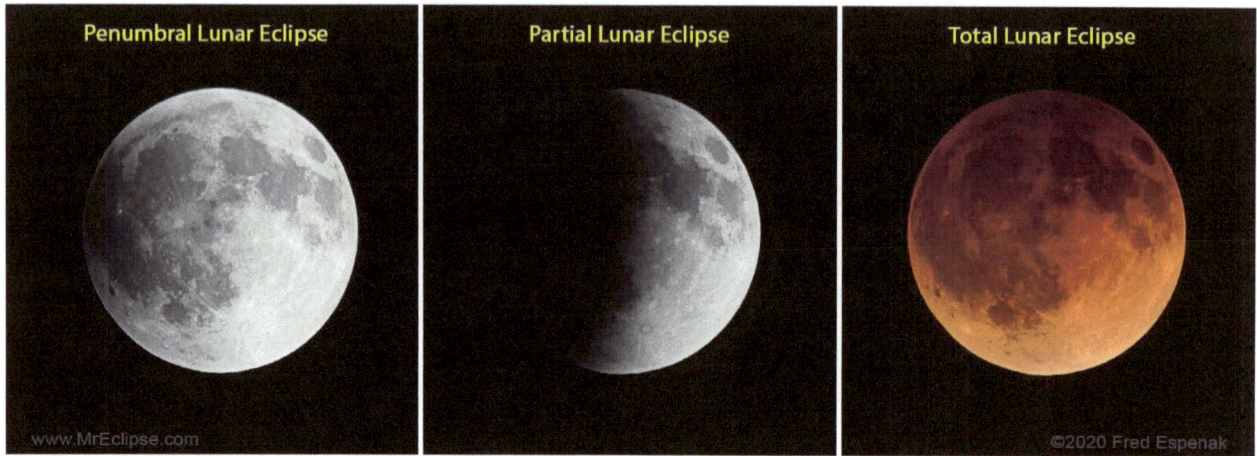

Photo 2–2 Examples of the visual appearance of a penumbral, partial, and total lunar eclipse. ©2020 F. Espenak

Types of Lunar Eclipses

There are three types of lunar eclipses:

1. **Penumbral Lunar Eclipse** — The Moon passes through Earth's faint penumbral shadow. Penumbral eclipses are of minor interest since they are quite subtle and difficult to observe.
2. **Partial Lunar Eclipse** — A portion of the Moon passes through Earth's dark umbral shadow. The remaining part of the Moon appears bright even though it lies deep within the penumbra. Partial eclipses are easy to see, even with the unaided eye.
3. **Total Lunar Eclipse** — The entire Moon passes through Earth's umbral shadow. Total eclipses are quite striking for the vibrant range of color of the Moon during the total phase, referred to as totality.

Figure 1–2. Types of Lunar Eclipses

Figure 2–2 illustrates the three types of lunar eclipses as seen from Earth. 1) A penumbral eclipse occurs when the Moon passes through the penumbra but completely misses the umbra. 2) A partial eclipse occurs if some portion of the Moon enters the umbra. 3) A total lunar eclipse takes place when the entire disk of the Moon enters the umbral shadow.

Photo 2–3 Penumbral lunar eclipse of 2002 Nov 20. Left: Moon before the eclipse begins. Right: mid-eclipse when 88.9% of the Moon's is in the penumbral shadow. ©2002 F. Espenak

Visual Appearance of Penumbral Lunar Eclipses

The visual appearance of penumbral, partial and total lunar eclipses differs significantly. While penumbral eclipses are pale and difficult to see, partial eclipses are easy naked-eye events while total eclipses are colorful and dramatic.

Earth's penumbral shadow forms a diverging cone that expands into space away from the Sun. Within this zone, Earth blocks part but not all of the Sun's disk. Some portion of the Sun's direct rays continue to reach the Moon during a penumbral eclipse.

The early and late stages of a penumbral eclipse are completely invisible to the eye. It is only when about 2/3 of the Moon's disk has entered the penumbral shadow that a skilled observer can detect a faint shading across the Moon.

Even when 90% of the Moon is immersed in the penumbra, approximately 10% of the Sun's rays still reach the Moon's deepest limb. Under such conditions, the Moon remains relatively bright with only a subtle shadow gradient across its disk.

Photo 2–4 Time sequence of the partial lunar eclipse of 2012 Jun 04. ©2012 F. Espenak

Visual Appearance of Partial Lunar Eclipses

Compared to penumbral eclipses, partial eclipses are easy to see with the naked eye. The lunar limb extending into the umbral shadow appears very dark or even black. This is due to a contrast effect since the remaining portion of the Moon in the penumbra is hundreds of times brighter. Because the umbral shadow's diameter is about 2.7 times the Moon's diameter, it appears as though a semi-circular bite has been taken out of the Moon.

Aristotle (384–322 BCE) first proved that Earth was round using the curved umbral shadow seen at partial eclipses. In comparing observations of several eclipses, he noted that Earth's shadow was round no matter where the eclipse took place, whether the Moon was high in the sky or low near the horizon. Aristotle reasoned that only a sphere casts a round shadow from every angle.

Photo 2–5 The beginning, middle and end of totality during the total lunar eclipse of 2004 October 28. ©2004 F. Espenak

Visual Appearance of Total Lunar Eclipses

A total lunar eclipse is the most dramatic and visually compelling type of lunar eclipse. The Moon's appearance can vary enormously throughout the period of totality and from one eclipse to the next. The geometry of the Moon's path through the umbra plays a significant role in determining the appearance of totality. The effect that Earth's atmosphere has on a total eclipse is not as apparent. Although the physical mass of Earth blocks all direct sunlight from the umbra, the planet's atmosphere filters, attenuates and bends some of the Sun's rays into the shadow.

The molecules in Earth's atmosphere scatter short wavelength light (i.e., yellow, green, blue) more than long wavelength light (i.e., orange, red). The same process responsible for making sunsets red also gives total lunar eclipses their characteristic ruddy color. The exact appearance can vary widely in both hue and brightness.

Because the lowest layers of the atmosphere are the thickest, they absorb more sunlight and refract it through larger angles. About 75% of the atmosphere's mass is concentrated in the bottom 10 kilometers (troposphere) as well as most of the water vapor, which can form massive clouds that block even more light. Just above the troposphere lies the stratosphere (10 to 50 kilometers), a rarified zone above most of the planet's weather systems. The stratosphere is subject to important photochemical reactions due to the high level of solar ultraviolet radiation that penetrates the region. The troposphere and stratosphere act together as a ring-shaped lens that refracts heavily reddened sunlight into Earth's umbral shadow. Since the higher stratospheric layers contain less gas, they refract sunlight through progressively smaller angles into the outer parts of the umbra. In contrast, denser tropospheric layers refract sunlight through larger angles to reach the inner parts of the umbra.

Because of this lensing effect, the amount of light refracted into the umbra tends to decrease radially from the edge to the center. Inhomogeneities from varying amounts of cloud and dust at differing latitudes can cause significant variations in brightness throughout the umbra.

30

Besides water (cloud, mist, precipitation), Earth's atmosphere also contains aerosols or tiny particles of organic debris, meteoric dust, volcanic ash and photochemical droplets. This material attenuates sunlight before it is refracted into the umbra. For instance, major volcanic eruptions in 1963 (Agung) and 1982 (El Chichon) each dumped large quantities of gas and ash into the stratosphere and were followed by several years of dark eclipses.

The 1991 eruption of Pinatubo in the Philippines had a similar effect. While most of the solid ash fell to Earth several days after circulating through the troposphere, a sizable volume of sulfur dioxide (SO_2) reached the stratosphere where it interacts with water vapor and eventually produces sulfuric acid (H_2SO_4). This high-altitude volcanic haze layer severely attenuates sunlight that must travel several hundred kilometers horizontally through the layer before being refracted into the umbral shadow. Thus, total eclipses following large volcanic eruptions are particularly dark. The total lunar eclipse of 1992 Dec 09 (1½ years after Pinatubo) was so dark that it was difficult to see the Moon's dull gray disk with the naked eye.

All total eclipses begin with penumbral and partial phases. After the total phase, the eclipse ends with more partial and penumbral phases. Lunar eclipses are completely safe to view and require none of the precautions needed for viewing solar eclipses (like special filters). The best views of a lunar eclipse are with binoculars and the naked eye.

Figure 2–1. Lunar Eclipse Contacts

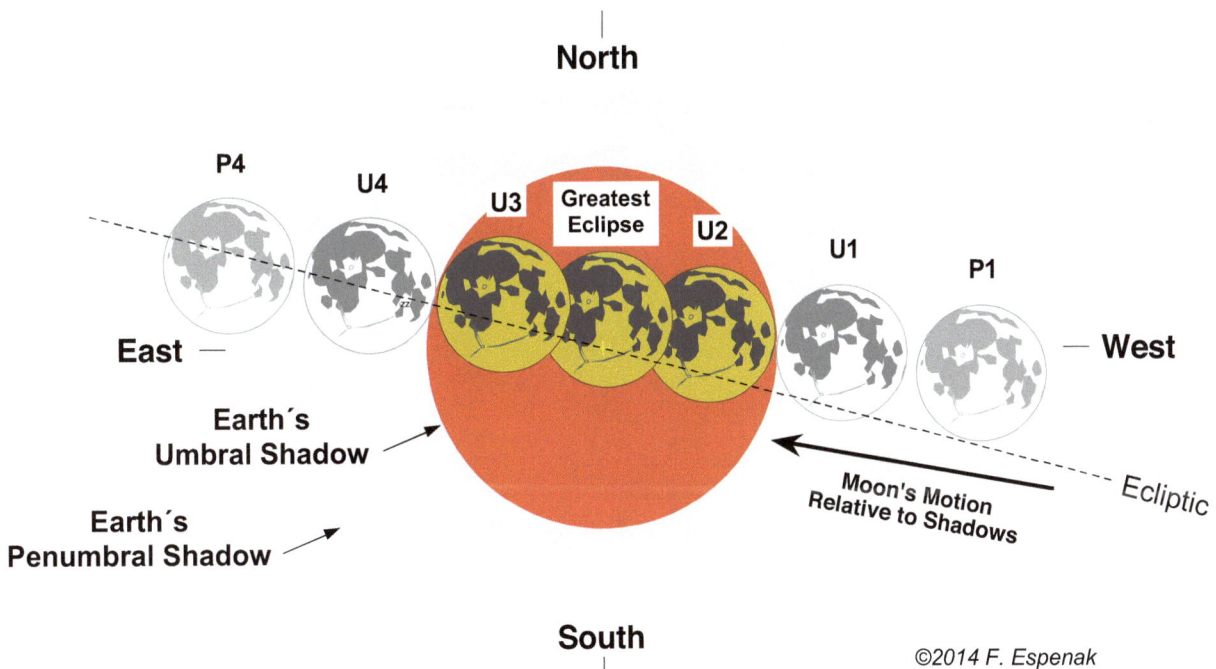

©2014 F. Espenak

Figure 2–3 illustrates the six contacts for a total lunar eclipse. These correspond to the instants when the Moon's disk is externally tangent to the penumbra (P1 and P4), or either externally or internally tangent to the umbra (U1, U2, U3, and U4). Partial eclipses do not have contacts U2 and U3, while penumbral eclipses only have contacts P1 and P4.

Lunar Eclipse Contacts

During the course of a lunar eclipse, the instants when the Moon's disk becomes tangent to Earth's shadows are known as eclipse contacts. They mark the primary stages or phases of a lunar eclipse (see figure 2-3) and are defined as follows.

P1 — Penumbral Eclipse Begins	(Instant of first exterior tangency of the Moon with the Penumbra)
U1 — Partial Eclipse Begins	(Instant of first exterior tangency of the Moon with the Umbra)
U2 — Total Eclipse Begins	(Instant of first interior tangency of the Moon with the Umbra)
U3 — Total Eclipse Ends	(Instant of last interior tangency of the Moon with the Umbra)
U4 — Partial Eclipse Ends	(Instant of last exterior tangency of the Moon with the Umbra)
P4 — Penumbral Eclipse Ends	(Instant of last exterior tangency of the Moon with the Penumbra)

Penumbral eclipses only have first and last contacts (i.e., P1 and P4) although neither of these events is observable.

In addition to P1 and P4, partial eclipses also have contacts U1 and U4 when the partial phases begin and end.

Total lunar eclipses have all six contacts. Contacts U2 and U3 mark the instants when the Moon's entire disk is first and last internally tangent to the umbra. These are the times when the total phase of the eclipse begins and ends..

The instant of greatest eclipse occurs when the Moon passes closest to the shadow axis. This is the maximum phase of the eclipse when the Moon is at its deepest position within either the penumbral or umbral shadow.

Photo 2–6 Five minutes before the start of totality (2018 Jan 31), the Moon is bathed in an orange-red light. The narrow rim outside the umbra and still in sunlight appears brilliant white. ©2018 F. Espenak

Enlargement of Earth's Shadows

In 1707, Philippe de La Hire made a curious observation about Earth's umbra. The predicted radius of the shadow needed to be enlarged by about 1/41 in order to fit timings made during a lunar eclipse. The enlargement is attributed to Earth's atmosphere, which becomes increasingly less transparent at lower levels. Additional observations over the next two centuries revealed that the shadow enlargement was somewhat variable from one eclipse to the next.

William Chauvenet (1891) formulated one method to account for the shadow enlargement while André-Louis Danjon (1951) devised another. Both methods assumed a circular cross section for the umbral shadow. However, Earth is flattened at the poles and bulges at the Equator, so an oblate spheroid more closely represents its shape. The projection of each of the planet's shadows is an ellipse rather than a circle. Furthermore, Earth's axial tilt towards or away from the Sun throughout the year means the elliptical shape of the penumbral and umbral shadows varies as well.

In an analysis of 22,539 observations made at 94 lunar eclipses from 1842 to 2011, Herald and Sinnott[7] (2014) found that the size and shape of the umbra are consistent with an oblate spheroid at the time of each eclipse, enlarged by the empirically determined occulting layer that uniformly surrounds Earth. The effective height of this layer was found to be 87 kilometers. Based on this work, the authors developed a new method to calculate the shadow enlargement including their elliptical shape.

This new method is the most rigorous and accurate procedure to date. The *Eclipse Almanac* uses it in the lunar eclipse predictions presented here.

Explanation of Lunar Eclipse Figures

There are 23 eclipses of the Moon during the period 2051 to 2060. A figure for each eclipse is included.

Each figure consists of two diagrams. The first one depicts the Moon's path through Earth's penumbral and umbral shadows (with Celestial North up). The second is a map showing the geographic visibility of each eclipse phase. All features in these diagrams are identified in the key on the next page.

The Moon's orbital motion with respect to the shadows is from west to east (right to left). Each phase of the eclipse is defined by the instant when the Moon's limb is externally or internally tangent to the penumbra or umbra. The six primary contacts of the Moon with the penumbral and umbral shadows are defined as follows.

P1 — Penumbral Eclipse Begins (Instant of first exterior tangency of the Moon with the Penumbra)
U1 — Partial Eclipse Begins (Instant of first exterior tangency of the Moon with the Umbra)
U2 — Total Eclipse Begins (Instant of first interior tangency of the Moon with the Umbra)
U3 — Total Eclipse Ends (Instant of last interior tangency of the Moon with the Umbra)
U4 — Partial Eclipse Ends (Instant of last exterior tangency of the Moon with the Umbra)
P4 — Penumbral Eclipse Ends (Instant of last exterior tangency of the Moon with the Penumbra)

Penumbral lunar eclipses have two primary contacts: P1 and P4, but neither is observable because the edge of the penumbra is indistinct and extremely faint.

In addition to the penumbral contacts, partial lunar eclipses have two more contacts as the Moon's limb enters and exits the umbral shadow: U1 and U4, respectively. These two contacts mark the instants when the partial phase of the eclipse begins and ends.

Total lunar eclipses undergo all six contacts. The two additional umbral contacts are the instants when the Moon's entire disk is first and last internally tangent to the umbra: U2 and U3, respectively. They mark the times when the total phase of the eclipse begins and ends.

The Moon passes closest to the shadow axis at the instant of greatest eclipse. This corresponds to the maximum phase of the eclipse and the Moon's position at this instant is also depicted in the path diagrams.

The equidistant cylindrical projection map shows the geographic region of visibility at each phase of the eclipse. This is accomplished using a series of curves showing where Moonrise and Moonset occur at each eclipse contact. The map is shaded to indicate eclipse visibility. The entire eclipse is visible from the zone with no shading. Conversely, none of the eclipse can be seen from the zone with the darkest shading.

At greatest eclipse, the Moon is deepest in Earth's shadow. The geographic location where the Moon appears in the zenith at greatest eclipse is shown by an asterisk.

Parameters relevant to the eclipse appear in on the right side of each figure. The instant of greatest eclipse is the time (Universal Time[8] or UT1) when the Moon passes closest to the shadow axis. The penumbral and umbral magnitudes are the fractions of the Moon's diameter immersed in each shadow. Gamma is the minimum distance of the Moon's center from the axis of Earth's shadow at greatest eclipse. The Saros series of the eclipse, and the

---

[7] Herald, D., and Sinnott, R. W., "Analysis of Lunar Crater Timings, 1842–2011," *J. Br. Astron. Assoc.*, **124**, 5 (2014)

[8] Universal Time (UT1) is the modern-day replacement for Greenwich Mean Time and is based on Earth's rotation with respect to distant quasars.

node of the Moon's orbit are given. Each contact time of the Moon's edge with the penumbral and umbral shadows is listed in Universal Time (UT1). Depending on the eclipse type, the duration of the penumbral, partial or total phases are given.

2–7 A time sequence shows the partial phases and totality during the total lunar eclipse of 2000 July 16 from Maui, Hawaii (Copyright ©2000 by Fred Espenak).

Lunar Eclipse Figures

Key to Lunar Eclipse Figures

Total Lunar Eclipse
2029 Jun 26

Greatest Eclipse = 03:22:08.8 UT1
Penumbral Eclipse Magnitude = 2.8333
Umbral Eclipse Magnitude = 1.8503
Gamma = 0.0124
Axis = 0.0121°
Saros = 130 [35 of 71]
Ascending Node

Geocentric Coordinates at Greatest Eclipse

Sun	Moon
R.A. = 06h21m03.1s	R.A. = 18h21m02.6s
Dec. = +23°20'50.2"	Dec. = -23°20'06.9"
S.D. = 00°15'44.1"	S.D. = 00°16'00.4"
H.P. = 00°00'08.7"	H.P. = 00°58'44.7"

Eclipse Durations	Eclipse Contacts
Penumbral = 05h36m02s	P1 = 00:34:11 UT1
Umbral = 03h40m21s	U1 = 01:31:57 UT1
Total = 01h42m41s	U2 = 02:30:48 UT1
	U3 = 04:13:29 UT1
	U4 = 05:12:18 UT1
Parameters	P4 = 06:10:13 UT1
Eph. = JPL DE430	
Rule = Herald-Sinnott	
ΔT = 74 s	

www.EclipseWise.com   ©2020 F. Espenak

Eclipse During Moon Rise
P1 - Penumbral Eclipse Begins
U1 - Partial Eclipse Begins
U2 - Total Eclipse Begins
U3 - Total Eclipse Ends
U4 - Partial Eclipse Ends
P4 - Penumbral Eclipse Ends

Eclipse During Moon Set
P1 - Penumbral Eclipse Begins
U1 - Partial Eclipse Begins
U2 - Total Eclipse Begins
U3 - Total Eclipse Ends
U4 - Partial Eclipse Ends
P4 - Penumbral Eclipse Ends

Layout of Lunar Eclipse Figure
Upper Left: Moon's Path Thru Earth's Shadows
Lower Left: Map of Eclipse Visibility
Right Side: Parameters for Lunar Eclipse

Explanation of Parameters Used in Lunar Eclipse Figures

Greatest Eclipse – The instant when the Moon passes closest to the axis of Earth's shadow cone (Universal Time[9])
Penumbral Eclipse Magnitude – Fraction of the Moon's diameter immersed in the penumbra at greatest eclipse.
Umbral Eclipse Magnitude – The fraction of the Moon's diameter immersed in the umbra at greatest eclipse.
Gamma – Minimum distance from the Moon's center to Earth's shadow axis (units of Earth's equatorial radius).
Axis – Minimum distance from the Moon's center to Earth's shadow axis (units of degrees).
Saros Series – The Saros series that the eclipse belongs to. The numbers in "[ ]" are the eclipse's sequential position and the number of eclipses in the Saros series.
Node – The orbital node near which the eclipse takes place (Ascending Node or Descending Node).
Geocentric Coordinates of the Sun and the Moon at Greatest Eclipse

    R.A. – Right Ascension   **S.D. –** Semi-Diameter (i.e. - radius)
    Dec. – Declination   **H.P. –** Horizontal Parallax

Eclipse Durations – Durations of Penumbral, Partial, and Total Eclipse.
Eclipse Contacts – Contact Times (Universal Time or UT1) of the Moon with the Penumbra and the Umbra
    P1, P4 – Start and End of the Penumbral Eclipse
    U1, U4 – Start and End of the Partial Eclipse
    U2, U3 – Start and End of the Total Eclipse

---

[9] Universal Time or UT1 is the modern replacement for Greenwich Mean Time

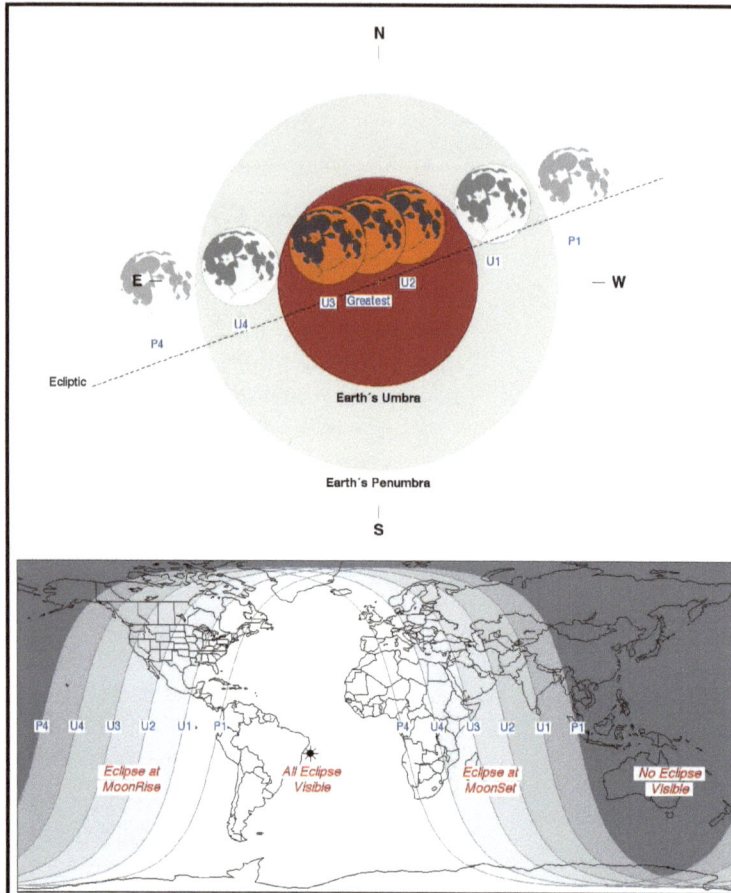

Total Lunar Eclipse
2051 Apr 26

Greatest Eclipse = 02:15:01.8 UT1
Penumbral Eclipse Magnitude = 2.2785
Umbral Eclipse Magnitude = 1.2034
Gamma = 0.3371
Axis = 0.3049°
Saros = 132 [32 of 71]
Ascending Node

Geocentric Coordinates at Greatest Eclipse

Sun	Moon
R.A. = 02h14m06.4s	R.A. = 14h14m24.0s
Dec. = +13°27'39.8"	Dec. = -13°09'52.9"
S.D. = 00°15'53.9"	S.D. = 00°14'47.3"
H.P. = 00°00'08.7"	H.P. = 00°54'16.4"

Eclipse Durations	Eclipse Contacts
Penumbral = 06h05m43s	P1 = 23:12:11 UT1
Umbral = 03h41m40s	U1 = 00:24:16 UT1
Total = 01h10m15s	U2 = 01:40:03 UT1
	U3 = 02:50:18 UT1
	U4 = 04:05:56 UT1
Parameters	P4 = 05:17:54 UT1
Eph. = JPL DE430	
Rule = Herald-Sinnott	
ΔT = 86 s	

www.EclipseWise.com   ©2020 F. Espenak

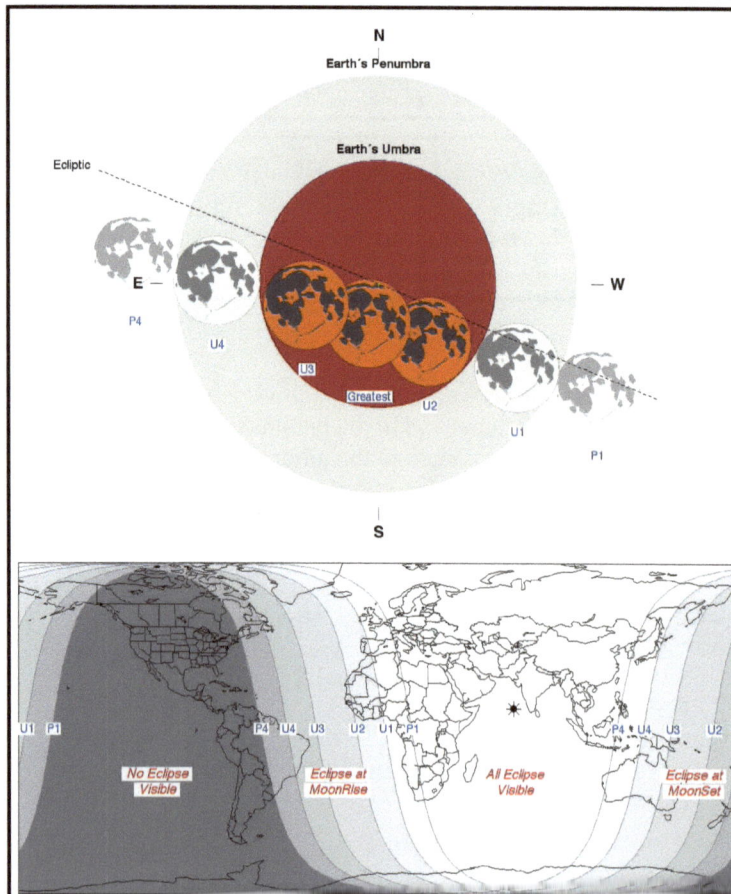

Total Lunar Eclipse
2051 Oct 19

Greatest Eclipse = 19:10:23.3 UT1
Penumbral Eclipse Magnitude = 2.3719
Umbral Eclipse Magnitude = 1.4130
Gamma = -0.2542
Axis = 0.2603°
Saros = 137 [28 of 78]
Decending Node

Geocentric Coordinates at Greatest Eclipse

Sun	Moon
R.A. = 13h37m47.1s	R.A. = 01h38m04.1s
Dec. = -10°10'03.3"	Dec. = +09°55'00.4"
S.D. = 00°16'03.4"	S.D. = 00°16'44.6"
H.P. = 00°00'08.8"	H.P. = 01°01'27.1"

Eclipse Durations	Eclipse Contacts
Penumbral = 05h14m59s	P1 = 16:32:55 UT1
Umbral = 03h25m00s	U1 = 17:27:56 UT1
Total = 01h24m12s	U2 = 18:28:23 UT1
	U3 = 19:52:35 UT1
	U4 = 20:52:56 UT1
Parameters	P4 = 21:47:55 UT1
Eph. = JPL DE430	
Rule = Herald-Sinnott	
ΔT = 86 s	

www.EclipseWise.com   ©2020 F. Espenak

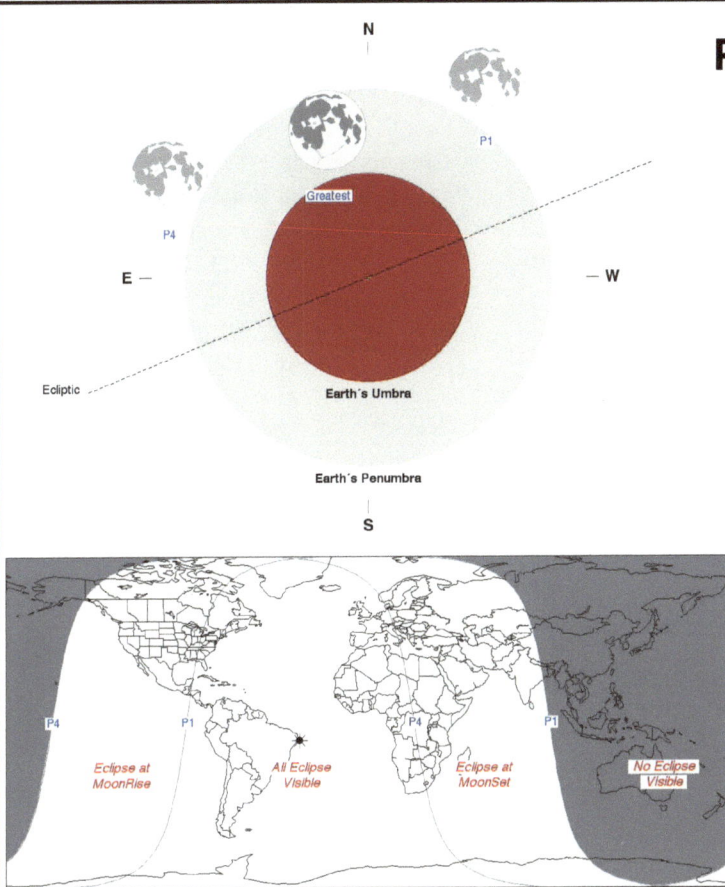

Penumbral Lunar Eclipse
2052 Apr 14

Greatest Eclipse = 02:16:39.7 UT1

Penumbral Eclipse Magnitude = 0.9478

Umbral Eclipse Magnitude = -0.1294

Gamma = 1.0629

Axis = 0.9625°

Saros = 142 [20 of 73]

Ascending Node

Geocentric Coordinates at Greatest Eclipse

Sun	Moon
R.A. = 01h32m05.6s	R.A. = 13h33m09.3s
Dec. = +09°37'10.9"	Dec. = -08°41'36.6"
S.D. = 00°15'56.9"	S.D. = 00°14'48.3"
H.P. = 00°00'08.8"	H.P. = 00°54'20.2"

Eclipse Durations	Eclipse Contacts
Penumbral = 04h36m46s	P1 = 23:58:23 UT1
	P4 = 04:35:09 UT1

Parameters
Eph. = JPL DE430

Rule = Herald-Sinnott

ΔT = 87 s

www.EclipseWise.com   ©2020 F. Espenak

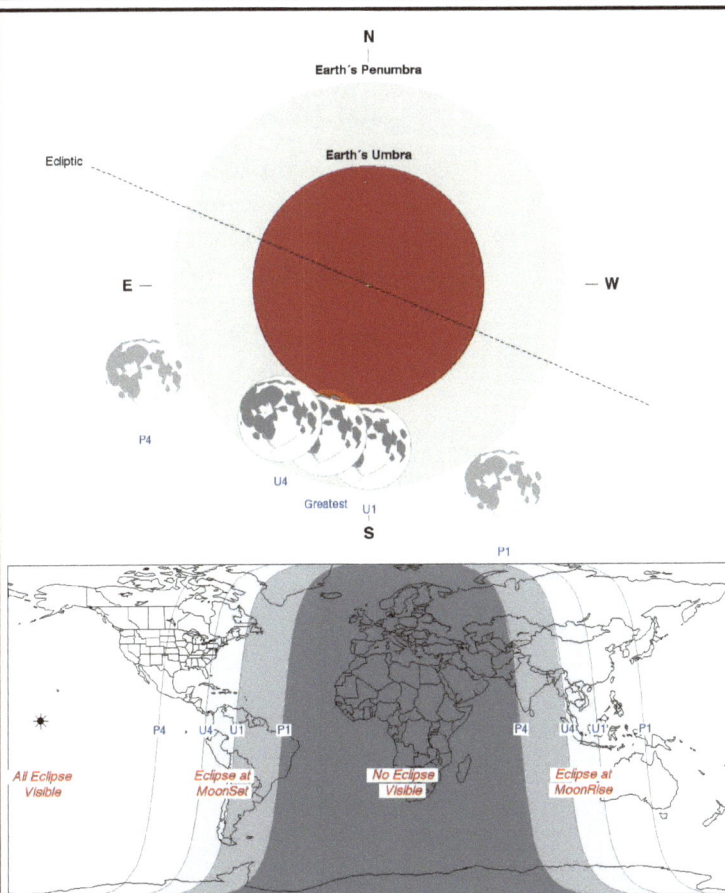

Partial Lunar Eclipse
2052 Oct 08

Greatest Eclipse = 10:44:31.1 UT1

Penumbral Eclipse Magnitude = 1.0653

Umbral Eclipse Magnitude = 0.0832

Gamma = -0.9727

Axis = 0.9697°

Saros = 147 [10 of 70]

Decending Node

Geocentric Coordinates at Greatest Eclipse

Sun	Moon
R.A. = 12h58m28.0s	R.A. = 00h59m36.6s
Dec. = -06°14'27.6"	Dec. = +05°18'49.9"
S.D. = 00°16'00.5"	S.D. = 00°16'18.0"
H.P. = 00°00'08.8"	H.P. = 00°59'49.3"

Eclipse Durations	Eclipse Contacts
Penumbral = 04h17m19s	P1 = 08:36:01 UT1
Umbral = 01h03m54s	U1 = 10:12:50 UT1
	U4 = 11:16:44 UT1
	P4 = 12:53:20 UT1

Parameters
Eph. = JPL DE430

Rule = Herald-Sinnott

ΔT = 87 s

www.EclipseWise.com   ©2020 F. Espenak

Penumbral Lunar Eclipse
2053 Mar 04

Greatest Eclipse = 17:20:42.1 UT1

Penumbral Eclipse Magnitude = 0.9334

Umbral Eclipse Magnitude = -0.0796

Gamma = -1.0531

Axis = 1.0255°

Saros = 114 [61 of 71]

Ascending Node

Geocentric Coordinates at Greatest Eclipse

Sun	Moon
R.A. = 23h03m14.8s	R.A. = 11h02m02.1s
Dec. = -06°03'47.9"	Dec. = +05°04'58.9"
S.D. = 00°16'07.7"	S.D. = 00°15'55.3"
H.P. = 00°00'08.9"	H.P. = 00°58'26.0"

Eclipse Durations	Eclipse Contacts
Penumbral = 04h11m47s	P1 = 15:14:37 UT1
	P4 = 19:26:25 UT1

Parameters
Eph. = JPL DE430

Rule = Herald-Sinnott

ΔT = 87 s

www.EclipseWise.com   ©2020 F. Espenak

Penumbral Lunar Eclipse
2053 Aug 29

Greatest Eclipse = 08:04:22.3 UT1

Penumbral Eclipse Magnitude = 1.0203

Umbral Eclipse Magnitude = -0.0319

Gamma = 1.0165

Axis = 0.9358°

Saros = 119 [63 of 82]

Decending Node

Geocentric Coordinates at Greatest Eclipse

Sun	Moon
R.A. = 10h32m52.4s	R.A. = 22h31m49.7s
Dec. = +09°08'07.1"	Dec. = -08°14'09.0"
S.D. = 00°15'50.2"	S.D. = 00°15'03.1"
H.P. = 00°00'08.7"	H.P. = 00°55'14.5"

Eclipse Durations	Eclipse Contacts
Penumbral = 04h38m31s	P1 = 05:45:01 UT1
	P4 = 10:23:32 UT1

Parameters
Eph. = JPL DE430

Rule = Herald-Sinnott

ΔT = 87 s

www.EclipseWise.com   ©2020 F. Espenak

Total Lunar Eclipse
2054 Feb 22

Greatest Eclipse = 06:49:59.1 UT1
Penumbral Eclipse Magnitude = 2.2502
Umbral Eclipse Magnitude = 1.2780
Gamma = -0.3242
Axis = 0.3298°
Saros = 124 [51 of 73]
Ascending Node

Geocentric Coordinates at Greatest Eclipse

Sun	Moon
R.A. = 22h23m02.4s	R.A. = 10h22m40.9s
Dec. = -10°05'18.4"	Dec. = +09°46'14.8"
S.D. = 00°16'10.2"	S.D. = 00°16'38.0"
H.P. = 00°00'08.9"	H.P. = 01°01'02.7"

Eclipse Durations	Eclipse Contacts
Penumbral = 05h15m33s	P1 = 04:12:11 UT1
Umbral = 03h21m36s	U1 = 05:09:07 UT1
Total = 01h12m44s	U2 = 06:13:29 UT1
	U3 = 07:26:14 UT1
	U4 = 08:30:43 UT1
	P4 = 09:27:44 UT1

Parameters
Eph. = JPL DE430
Rule = Herald-Sinnott
ΔT = 88 s

www.EclipseWise.com   ©2020 F. Espenak

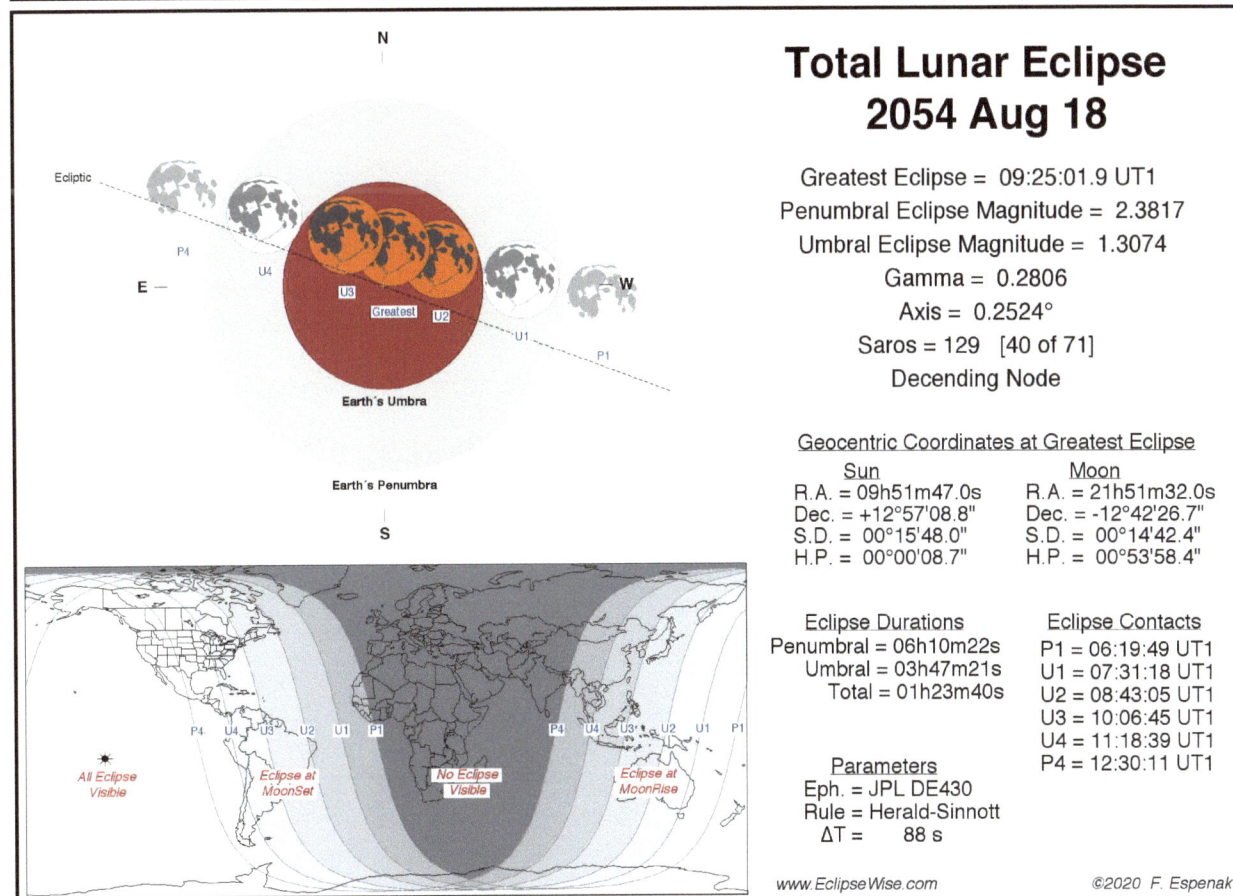

Total Lunar Eclipse
2054 Aug 18

Greatest Eclipse = 09:25:01.9 UT1
Penumbral Eclipse Magnitude = 2.3817
Umbral Eclipse Magnitude = 1.3074
Gamma = 0.2806
Axis = 0.2524°
Saros = 129 [40 of 71]
Decending Node

Geocentric Coordinates at Greatest Eclipse

Sun	Moon
R.A. = 09h51m47.0s	R.A. = 21h51m32.0s
Dec. = +12°57'08.8"	Dec. = -12°42'26.7"
S.D. = 00°15'48.0"	S.D. = 00°14'42.4"
H.P. = 00°00'08.7"	H.P. = 00°53'58.4"

Eclipse Durations	Eclipse Contacts
Penumbral = 06h10m22s	P1 = 06:19:49 UT1
Umbral = 03h47m21s	U1 = 07:31:18 UT1
Total = 01h23m40s	U2 = 08:43:05 UT1
	U3 = 10:06:45 UT1
	U4 = 11:18:39 UT1
	P4 = 12:30:11 UT1

Parameters
Eph. = JPL DE430
Rule = Herald-Sinnott
ΔT = 88 s

www.EclipseWise.com   ©2020 F. Espenak

Total Lunar Eclipse
2055 Feb 11

Greatest Eclipse = 22:44:49.0 UT1
Penumbral Eclipse Magnitude = 2.1982
Umbral Eclipse Magnitude = 1.2258
Gamma = 0.3526
Axis = 0.3594°
Saros = 134 [29 of 72]
Ascending Node

Geocentric Coordinates at Greatest Eclipse

Sun	Moon
R.A. = 21h42m03.5s	R.A. = 09h42m24.2s
Dec. = -13°47'10.8"	Dec. = +14°08'09.1"
S.D. = 00°16'12.3"	S.D. = 00°16'39.9"
H.P. = 00°00'08.9"	H.P. = 01°01'09.9"

Eclipse Durations	Eclipse Contacts
Penumbral = 05h13m41s	P1 = 20:08:00 UT1
Umbral = 03h19m09s	U1 = 21:05:18 UT1
Total = 01h06m35s	U2 = 22:11:39 UT1
	U3 = 23:18:14 UT1
	U4 = 00:24:27 UT1
Parameters	P4 = 01:21:40 UT1
Eph. = JPL DE430	
Rule = Herald-Sinnott	
ΔT = 88 s	

www.EclipseWise.com   ©2020 F. Espenak

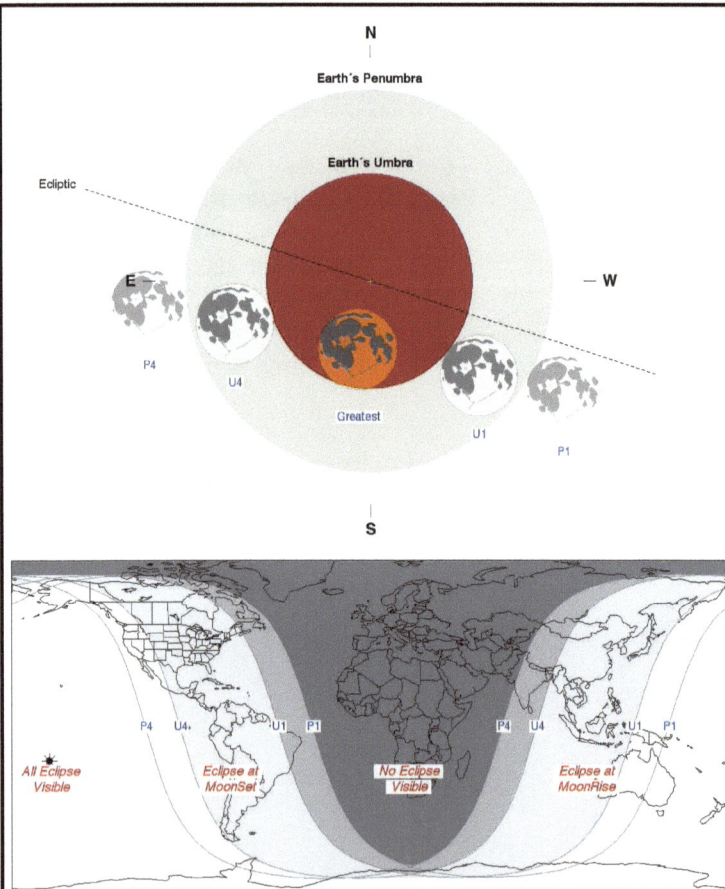

Partial Lunar Eclipse
2055 Aug 07

Greatest Eclipse = 10:51:49.4 UT1
Penumbral Eclipse Magnitude = 2.0081
Umbral Eclipse Magnitude = 0.9606
Gamma = -0.4769
Axis = 0.4391°
Saros = 139 [23 of 79]
Decending Node

Geocentric Coordinates at Greatest Eclipse

Sun	Moon
R.A. = 09h09m39.9s	R.A. = 21h10m01.5s
Dec. = +16°20'36.9"	Dec. = -16°46'26.9"
S.D. = 00°15'46.2"	S.D. = 00°15'03.3"
H.P. = 00°00'08.7"	H.P. = 00°55'15.1"

Eclipse Durations	Eclipse Contacts
Penumbral = 05h47m14s	P1 = 07:58:17 UT1
Umbral = 03h24m10s	U1 = 09:09:47 UT1
	U4 = 12:33:56 UT1
	P4 = 13:45:31 UT1
Parameters	
Eph. = JPL DE430	
Rule = Herald-Sinnott	
ΔT = 89 s	

www.EclipseWise.com   ©2020 F. Espenak

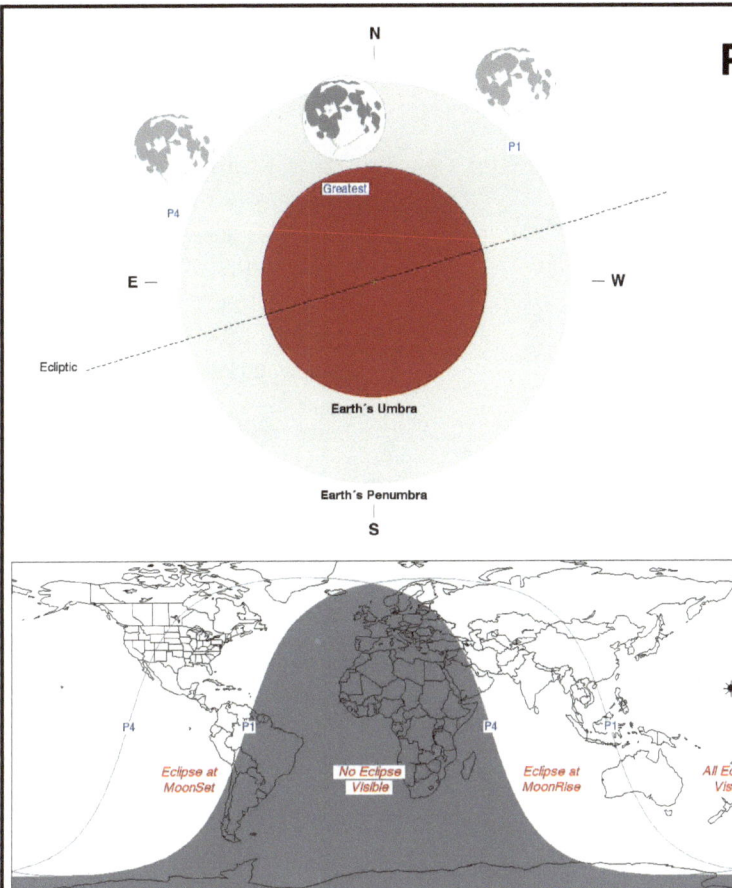

Penumbral Lunar Eclipse
2056 Feb 01

Greatest Eclipse = 12:24:37.1 UT1
Penumbral Eclipse Magnitude = 0.9069
Umbral Eclipse Magnitude = -0.1084
Gamma = 1.0682
Axis = 1.0447°
Saros = 144 [18 of 71]
Ascending Node

Geocentric Coordinates at Greatest Eclipse

Sun	Moon
R.A. = 20h59m18.0s	R.A. = 09h00m07.0s
Dec. = -17°05'10.5"	Dec. = +18°06'45.8"
S.D. = 00°16'14.0"	S.D. = 00°15'59.4"
H.P. = 00°00'08.9"	H.P. = 00°58'41.2"

Eclipse Durations	Eclipse Contacts
Penumbral = 04h07m59s	P1 = 10:20:46 UT1
	P4 = 14:28:44 UT1

Parameters
Eph. = JPL DE430
Rule = Herald-Sinnott
ΔT = 89 s

www.EclipseWise.com   ©2020 F. Espenak

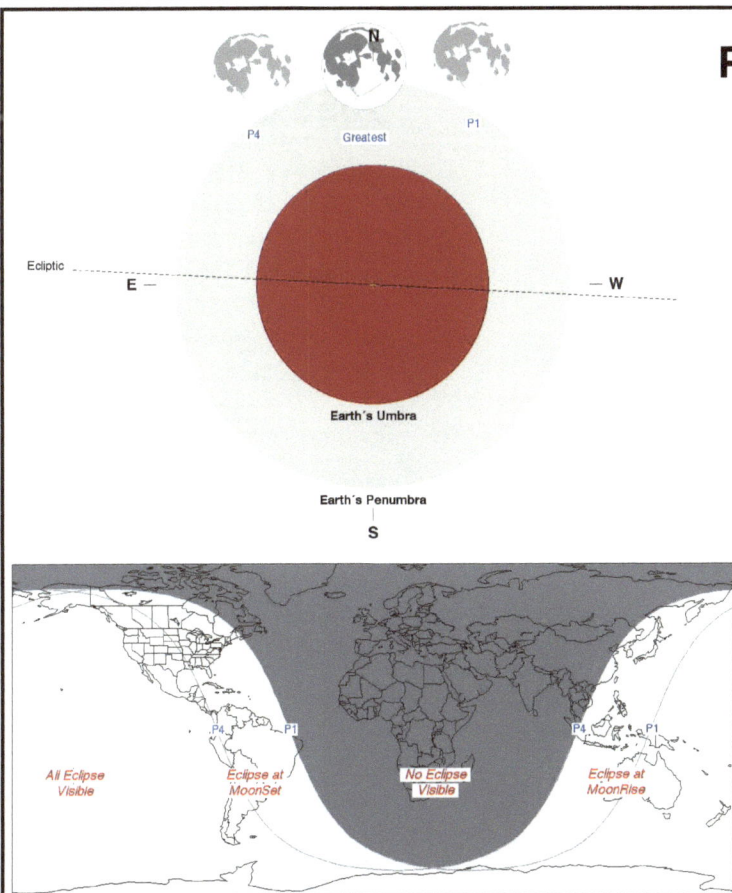

Penumbral Lunar Eclipse
2056 Jun 27

Greatest Eclipse = 10:01:39.9 UT1
Penumbral Eclipse Magnitude = 0.3158
Umbral Eclipse Magnitude = -0.6504
Gamma = 1.3770
Axis = 1.3714°
Saros = 111 [69 of 71]
Decending Node

Geocentric Coordinates at Greatest Eclipse

Sun	Moon
R.A. = 06h28m18.7s	R.A. = 18h28m35.0s
Dec. = +23°16'17.2"	Dec. = -21°54'04.7"
S.D. = 00°15'44.0"	S.D. = 00°16'17.0"
H.P. = 00°00'08.7"	H.P. = 00°59'45.8"

Eclipse Durations	Eclipse Contacts
Penumbral = 02h30m38s	P1 = 08:46:16 UT1
	P4 = 11:16:54 UT1

Parameters
Eph. = JPL DE430
Rule = Herald-Sinnott
ΔT = 89 s

www.EclipseWise.com   ©2020 F. Espenak

Penumbral Lunar Eclipse
2056 Jul 26

Greatest Eclipse = 18:41:55.4 UT1

Penumbral Eclipse Magnitude = 0.6448

Umbral Eclipse Magnitude = -0.3477

Gamma = -1.2048

Axis = 1.1693°

Saros = 149 [5 of 71]

Decending Node

Geocentric Coordinates at Greatest Eclipse

Sun	Moon
R.A. = 08h27m22.5s	R.A. = 20h28m05.5s
Dec. = +19°07'40.1"	Dec. = -20°17'05.8"
S.D. = 00°15'44.9"	S.D. = 00°15'52.1"
H.P. = 00°00'08.7"	H.P. = 00°58'14.4"

Eclipse Durations	Eclipse Contacts
Penumbral = 03h35m10s	P1 = 16:54:20 UT1
	P4 = 20:29:30 UT1

Parameters
Eph. = JPL DE430

Rule = Herald-Sinnott

ΔT = 89 s

www.EclipseWise.com   ©2020 F. Espenak

Penumbral Lunar Eclipse
2056 Dec 22

Greatest Eclipse = 01:47:26.2 UT1

Penumbral Eclipse Magnitude = 0.7872

Umbral Eclipse Magnitude = -0.3093

Gamma = -1.1560

Axis = 1.0483°

Saros = 116 [60 of 73]

Ascending Node

Geocentric Coordinates at Greatest Eclipse

Sun	Moon
R.A. = 18h04m03.3s	R.A. = 06h04m28.2s
Dec. = -23°25'40.9"	Dec. = +22°23'02.6"
S.D. = 00°16'15.5"	S.D. = 00°14'49.6"
H.P. = 00°00'08.9"	H.P. = 00°54'24.9"

Eclipse Durations	Eclipse Contacts
Penumbral = 04h17m16s	P1 = 23:38:53 UT1
	P4 = 03:56:09 UT1

Parameters
Eph. = JPL DE430

Rule = Herald-Sinnott

ΔT = 90 s

www.EclipseWise.com   ©2020 F. Espenak

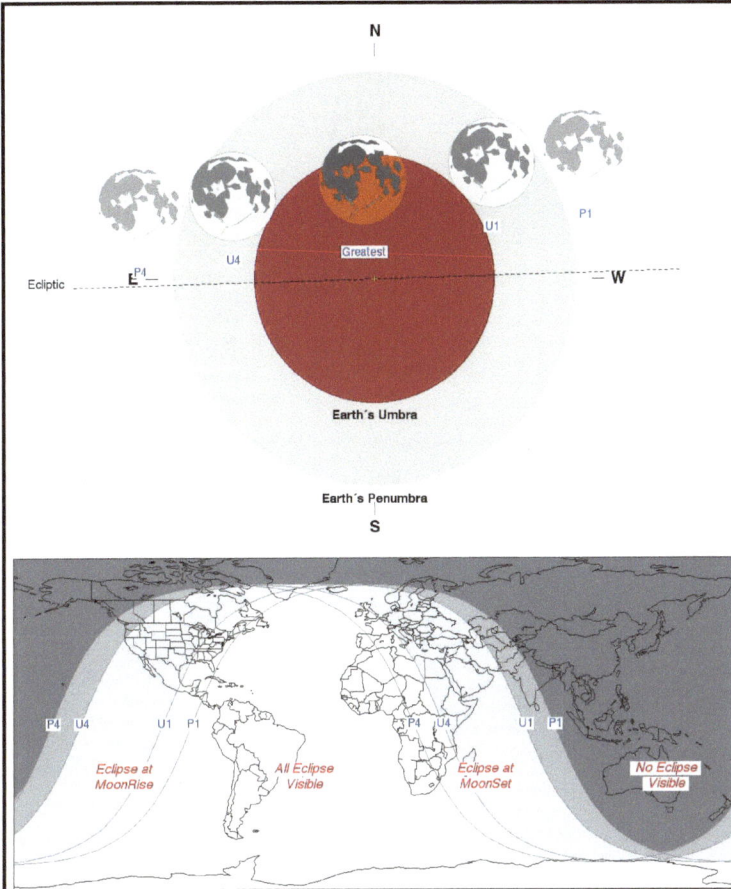

Partial Lunar Eclipse
2057 Jun 17

Greatest Eclipse = 02:24:50.0 UT1

Penumbral Eclipse Magnitude = 1.6982

Umbral Eclipse Magnitude = 0.7570

Gamma = 0.6168

Axis = 0.6310°

Saros = 121 [57 of 82]

Decending Node

Geocentric Coordinates at Greatest Eclipse

Sun	Moon
R.A. = 05h44m25.1s	R.A. = 17h44m45.6s
Dec. = +23°22'59.4"	Dec. = -22°45'25.3"
S.D. = 00°15'44.7"	S.D. = 00°16'43.7"
H.P. = 00°00'08.7"	H.P. = 01°01'23.6"

Eclipse Durations	Eclipse Contacts
Penumbral = 04h51m23s	P1 = 23:59:10 UT1
Umbral = 02h50m00s	U1 = 00:59:52 UT1
	U4 = 03:49:53 UT1
	P4 = 04:50:33 UT1

Parameters
Eph. = JPL DE430
Rule = Herald-Sinnott
ΔT = 90 s

www.EclipseWise.com   ©2020 F. Espenak

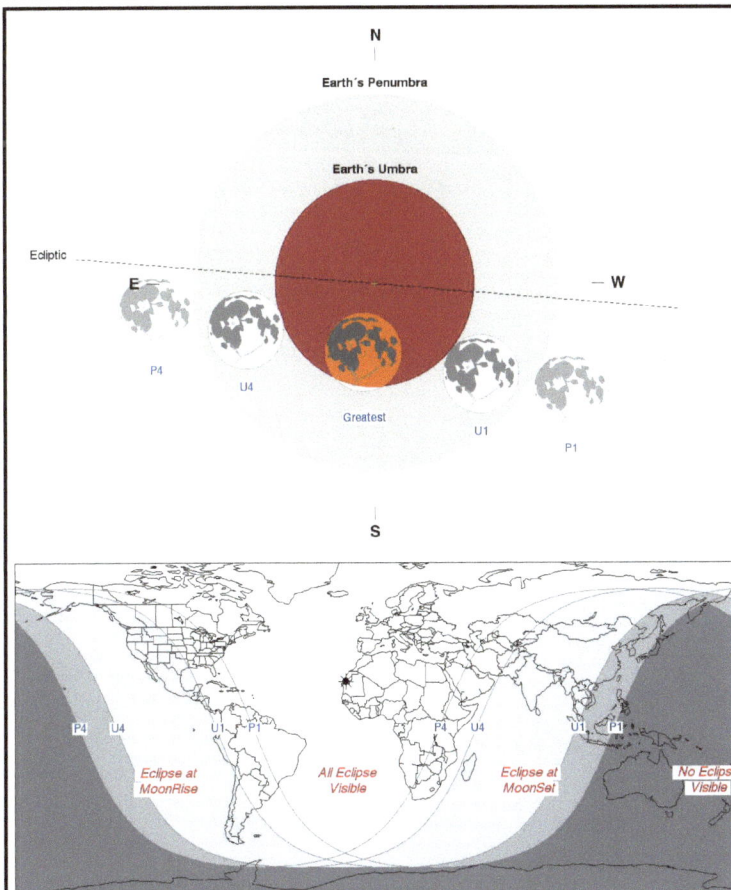

Partial Lunar Eclipse
2057 Dec 11

Greatest Eclipse = 00:52:07.6 UT1

Penumbral Eclipse Magnitude = 2.0194

Umbral Eclipse Magnitude = 0.9197

Gamma = -0.4853

Axis = 0.4384°

Saros = 126 [47 of 70]

Ascending Node

Geocentric Coordinates at Greatest Eclipse

Sun	Moon
R.A. = 17h14m07.3s	R.A. = 05h14m27.7s
Dec. = -23°00'46.9"	Dec. = +22°34'54.1"
S.D. = 00°16'14.5"	S.D. = 00°14'46.1"
H.P. = 00°00'08.9"	H.P. = 00°54'12.2"

Eclipse Durations	Eclipse Contacts
Penumbral = 05h59m42s	P1 = 21:52:20 UT1
Umbral = 03h24m50s	U1 = 23:09:45 UT1
	U4 = 02:34:35 UT1
	P4 = 03:52:02 UT1

Parameters
Eph. = JPL DE430
Rule = Herald-Sinnott
ΔT = 90 s

www.EclipseWise.com   ©2020 F. Espenak

Total Lunar Eclipse
2058 Jun 06

Greatest Eclipse = 19:14:17.3 UT1
Penumbral Eclipse Magnitude = 2.6226
Umbral Eclipse Magnitude = 1.6628
Gamma = -0.1181
Axis = 0.1186°
Saros = 131 [36 of 72]
Decending Node

Geocentric Coordinates at Greatest Eclipse

Sun	Moon
R.A. = 05h00m41.7s	R.A. = 17h00m35.5s
Dec. = +22°43'57.0"	Dec. = -22°50'55.4"
S.D. = 00°15'45.8"	S.D. = 00°16'25.3"
H.P. = 00°00'08.7"	H.P. = 01°00'16.2"

Eclipse Durations	Eclipse Contacts
Penumbral = 05h24m28s	P1 = 16:32:01 UT1
Umbral = 03h34m09s	U1 = 17:27:13 UT1
Total = 01h38m03s	U2 = 18:25:14 UT1
	U3 = 20:03:17 UT1
	U4 = 21:01:22 UT1
Parameters	P4 = 21:56:29 UT1
Eph. = JPL DE430	
Rule = Herald-Sinnott	
ΔT = 91 s	

www.EclipseWise.com   ©2020 F. Espenak

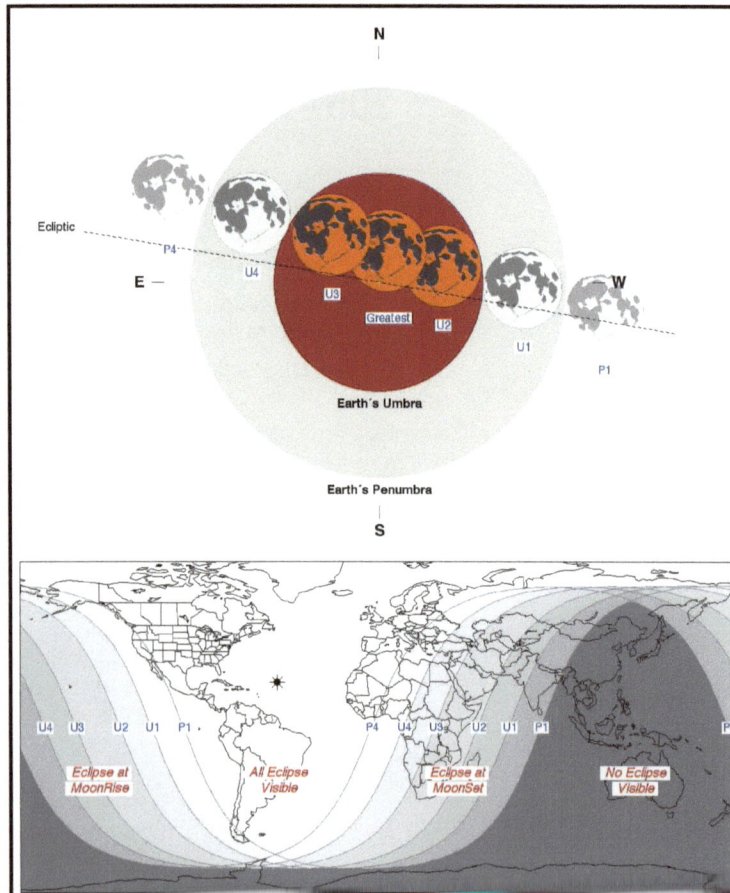

Total Lunar Eclipse
2058 Nov 30

Greatest Eclipse = 03:14:46.8 UT1
Penumbral Eclipse Magnitude = 2.4819
Umbral Eclipse Magnitude = 1.4277
Gamma = 0.2208
Axis = 0.2077°
Saros = 136 [22 of 72]
Ascending Node

Geocentric Coordinates at Greatest Eclipse

Sun	Moon
R.A. = 16h25m34.1s	R.A. = 04h25m20.3s
Dec. = -21°39'35.3"	Dec. = +21°51'37.8"
S.D. = 00°16'12.9"	S.D. = 00°15'22.8"
H.P. = 00°00'08.9"	H.P. = 00°56'26.8"

Eclipse Durations	Eclipse Contacts
Penumbral = 05h53m52s	P1 = 00:17:52 UT1
Umbral = 03h41m28s	U1 = 01:23:59 UT1
Total = 01h30m25s	U2 = 02:29:28 UT1
	U3 = 03:59:54 UT1
	U4 = 05:05:27 UT1
Parameters	P4 = 06:11:45 UT1
Eph. = JPL DE430	
Rule = Herald-Sinnott	
ΔT = 91 s	

www.EclipseWise.com   ©2020 F. Espenak

Partial Lunar Eclipse
2059 May 27

Greatest Eclipse = 07:54:02.8 UT1
Penumbral Eclipse Magnitude = 1.1963
Umbral Eclipse Magnitude = 0.1846
Gamma = -0.9098
Axis = 0.8684°
Saros = 141 [26 of 72]
Decending Node

Geocentric Coordinates at Greatest Eclipse

Sun	Moon
R.A. = 04h16m47.9s	R.A. = 16h15m47.5s
Dec. = +21°18'58.7"	Dec. = -22°09'09.5"
S.D. = 00°15'47.3"	S.D. = 00°15'36.4"
H.P. = 00°00'08.7"	H.P. = 00°57'16.6"

Eclipse Durations	Eclipse Contacts
Penumbral = 04h42m31s	P1 = 05:32:43 UT1
Umbral = 01h37m55s	U1 = 07:05:02 UT1
	U4 = 08:42:57 UT1
	P4 = 10:15:14 UT1

Parameters
Eph. = JPL DE430
Rule = Herald-Sinnott
ΔT = 91 s

www.EclipseWise.com   ©2020 F. Espenak

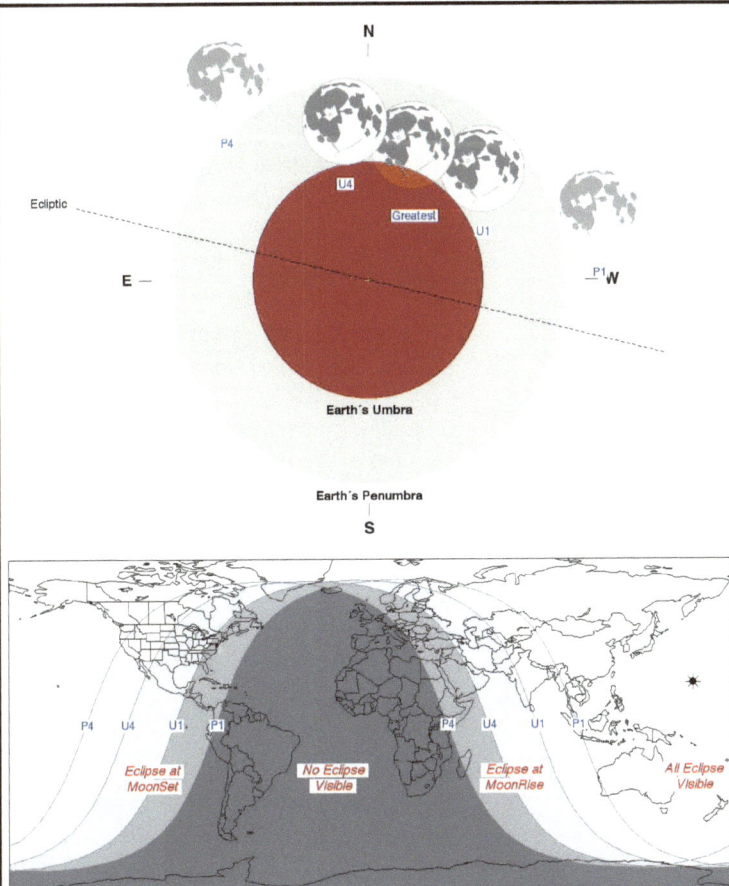

Partial Lunar Eclipse
2059 Nov 19

Greatest Eclipse = 13:00:03.8 UT1
Penumbral Eclipse Magnitude = 1.2055
Umbral Eclipse Magnitude = 0.2097
Gamma = 0.9004
Axis = 0.8949°
Saros = 146 [13 of 72]
Ascending Node

Geocentric Coordinates at Greatest Eclipse

Sun	Moon
R.A. = 15h39m38.2s	R.A. = 03h38m24.3s
Dec. = -19°31'36.5"	Dec. = +20°22'25.1"
S.D. = 00°16'10.9"	S.D. = 00°16'15.0"
H.P. = 00°00'08.9"	H.P. = 00°59'38.3"

Eclipse Durations	Eclipse Contacts
Penumbral = 04h31m15s	P1 = 10:44:18 UT1
Umbral = 01h39m56s	U1 = 12:09:51 UT1
	U4 = 13:49:47 UT1
	P4 = 15:15:33 UT1

Parameters
Eph. = JPL DE430
Rule = Herald-Sinnott
ΔT = 92 s

www.EclipseWise.com   ©2020 F. Espenak

Penumbral Lunar Eclipse
2060 Apr 15

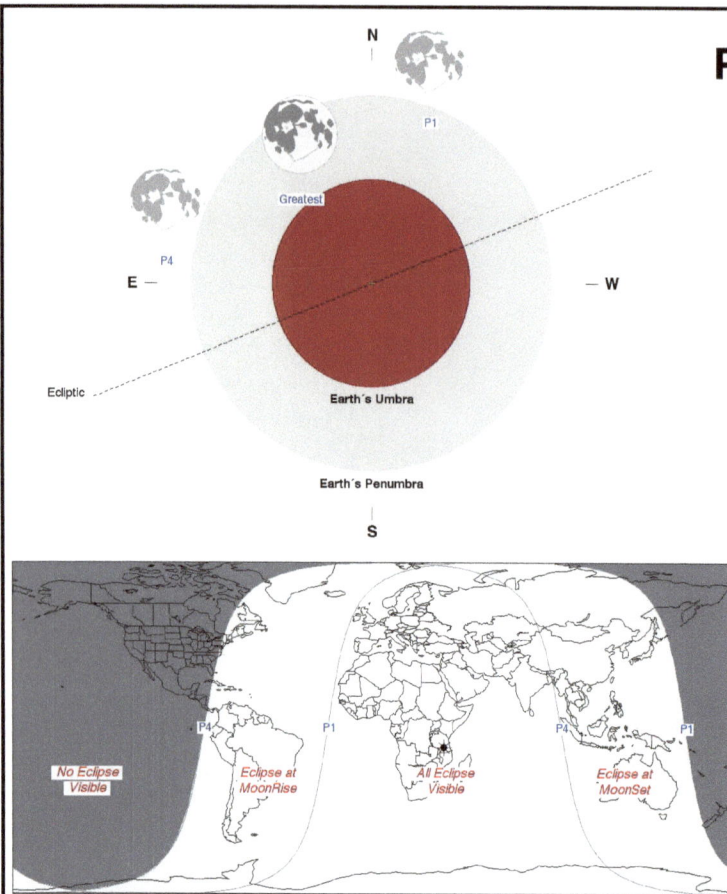

Greatest Eclipse = 21:35:32.8 UT1

Penumbral Eclipse Magnitude = 0.7694

Umbral Eclipse Magnitude = -0.3136

Gamma = 1.1622

Axis = 1.0462°

Saros = 113 [66 of 71]

Decending Node

Geocentric Coordinates at Greatest Eclipse

Sun	Moon
R.A. = 01h38m58.9s	R.A. = 13h40m53.7s
Dec. = +10°17'00.0"	Dec. = -09°20'57.2"
S.D. = 00°15'56.4"	S.D. = 00°14'43.1"
H.P. = 00°00'08.8"	H.P. = 00°54'01.1"

Eclipse Durations	Eclipse Contacts
Penumbral = 04h15m43s	P1 = 19:27:56 UT1
	P4 = 23:43:39 UT1

Parameters
Eph. = JPL DE430

Rule = Herald-Sinnott

ΔT = 92 s

Penumbral Lunar Eclipse
2060 Oct 09

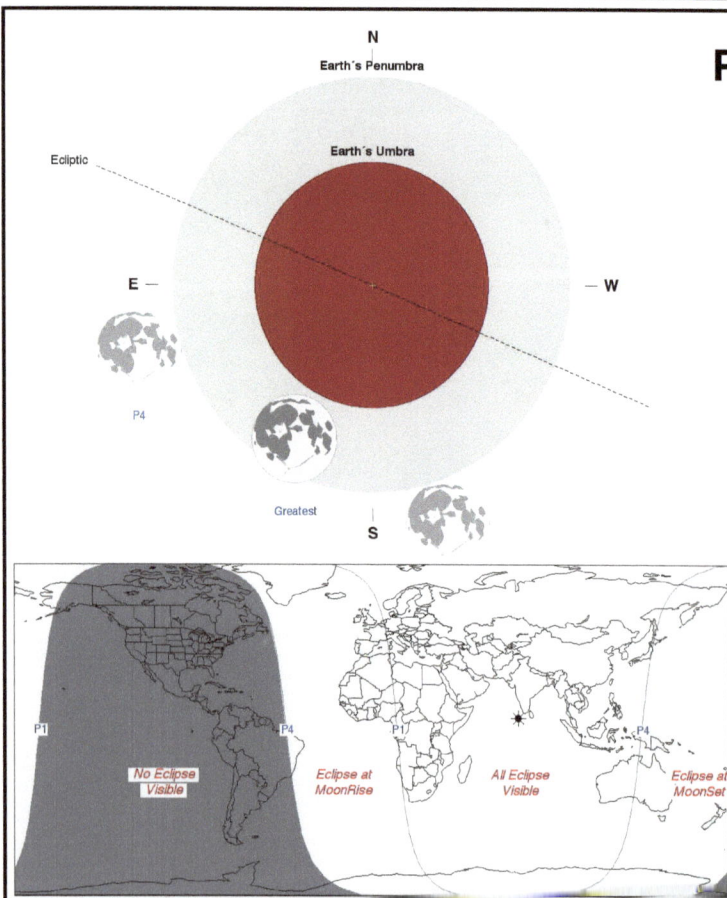

Greatest Eclipse = 18:52:00.2 UT1

Penumbral Eclipse Magnitude = 0.8816

Umbral Eclipse Magnitude = -0.0779

Gamma = -1.0671

Axis = 1.0892°

Saros = 118 [54 of 73]

Ascending Node

Geocentric Coordinates at Greatest Eclipse

Sun	Moon
R.A. = 13h03m35.8s	R.A. = 01h05m39.2s
Dec. = -06°46'17.7"	Dec. = +05°48'35.1"
S.D. = 00°16'00.8"	S.D. = 00°16'41.3"
H.P. = 00°00'08.8"	H.P. = 01°01'15.0"

Eclipse Durations	Eclipse Contacts
Penumbral = 03h52m00s	P1 = 16:56:13 UT1
	P4 = 20:48:13 UT1

Parameters
Eph. = JPL DE430

Rule = Herald-Sinnott

ΔT = 92 s

N

P4

P1

Greatest

Ecliptic

E —

— W

Earth's Umbra

Earth's Penumbra

S

Penumbral Lunar Eclipse
2060 Nov 08

Greatest Eclipse = 04:02:40.3 UT1

Penumbral Eclipse Magnitude = 0.0286

Umbral Eclipse Magnitude = -0.9356

Gamma = 1.5332

Axis = 1.5699°

Saros = 156 [1 of 81]

Ascending Node

Geocentric Coordinates at Greatest Eclipse

Sun	Moon
R.A. = 14h56m11.8s	R.A. = 02h53m43.2s
Dec. = -16°46'13.7"	Dec. = +18°13'31.2"
S.D. = 00°16'08.5"	S.D. = 00°16'44.5"
H.P. = 00°00'08.9"	H.P. = 01°01'26.6"

Eclipse Durations	Eclipse Contacts
Penumbral = 00h45m17s	P1 = 03:39:48 UT1
	P4 = 04:25:04 UT1

P4 P1

P4 P1

Eclipse at
MoonRise

All Eclipse
Visible

Eclipse at
MoonSet

No Eclipse
Visible

Parameters

Eph. = JPL DE430

Rule = Herald-Sinnott

ΔT =   93 s

www.EclipseWise.com

©2020 F. Espenak

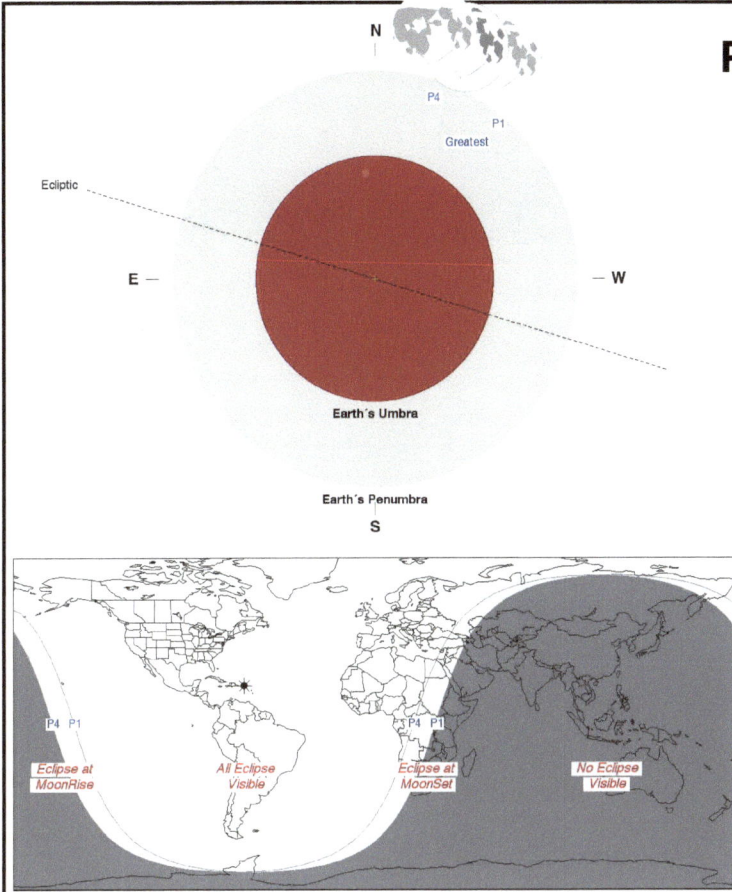

47

Key to Catalog of Lunar Eclipses

The following catalog lists all lunar eclipses during a 25-year period.

A brief description of each parameter in the catalog appears below.

Date — Gregorian date of Greatest Eclipse

Greatest Eclipse — Universal Time of Greatest Eclipse

Saros — Saros Series Number of the eclipse

Type — Lunar Eclipse Type

> N = Penumbral Lunar Eclipse
> P = Partial Lunar Eclipse
> T = Total Lunar Eclipse
>
> + = Central total eclipse (Moon's center passes north of shadow axis)
> − = Central total eclipse (Moon's center passes south of shadow axis)
>
> b = Saros series begins (first penumbral eclipse in a Saros series)
> e = Saros series ends (last penumbral eclipse in a Saros series)

Gamma — minimum distance from center of the Moon to axis of Earth's umbral shadow

P. Mag — Penumbral Magnitude; fraction of the Moon's diameter immersed in the penumbra

U. Mag — Umbral Magnitude; fraction of the Moon's diameter immersed in the umbra

Phase Durations
> **Pen** — elapsed time from contact P1 to P4 (minutes)
> **Partial** — elapsed time from contact U1 to U4 (minutes)
> **Total** — elapsed time from contact U2 to U3 (minutes)

Lat & Long — latitude and longitude where the Moon appears in the zenith at Greatest Eclipse

Photo 2–8 Time sequence of the total lunar eclipse of 2014 April 15. ©2014 F. Espenak

Catalog of Lunar Eclipses: 2051 to 2085

Date	Greatest Eclipse	Saros	Type	Gamma	P. Mag	U.Mag	Pen	Partial	Total	Lat.	Long.
2051 Apr 26	2:15:02	132	T	0.3371	2.2785	1.2034	365.7	221.7	70.3	13.2S	34.2W
2051 Oct 19	19:10:23	137	T-	-0.2542	2.3719	1.4130	315.0	205.0	84.2	9.9N	68.7E
2052 Apr 14	2:16:40	142	N	1.0629	0.9478	-0.1294	276.8	-	-	8.7S	33.8W
2052 Oct 08	10:44:31	147	P	-0.9727	1.0653	0.0832	257.3	63.9	-	5.3N	164.0W
2053 Mar 04	17:20:42	114	N	-1.0531	0.9334	-0.0796	251.8	-	-	5.1N	102.4E
2053 Aug 29	8:04:22	119	Nx	1.0165	1.0203	-0.0319	278.5	-	-	8.2S	121.1W
2054 Feb 22	6:49:59	124	T	-0.3242	2.2502	1.2780	315.5	201.6	72.7	9.8N	99.2W
2054 Aug 18	9:25:02	129	T	0.2806	2.3817	1.3074	370.4	227.4	83.7	12.7S	140.4W
2055 Feb 11	22:44:49	134	T	0.3526	2.1982	1.2258	313.7	199.1	66.6	14.1N	22.4E
2055 Aug 07	10:51:49	139	P	-0.4769	2.0081	0.9606	347.2	204.2	-	16.8S	161.4W
2056 Feb 01	12:24:37	144	N	1.0682	0.9069	-0.1084	248.0	-	-	18.1N	177.4E
2056 Jun 27	10:01:40	111	N	1.3770	0.3158	-0.6504	150.6	-	-	21.9S	149.5W
2056 Jul 26	18:41:55	149	N	-1.2048	0.6448	-0.3477	215.2	-	-	20.3S	81.3E
2056 Dec 22	1:47:26	116	N	-1.1560	0.7872	-0.3093	257.3	-	-	22.4N	27.1W
2057 Jun 17	2:24:50	121	P	0.6168	1.6982	0.7570	291.4	170.0	-	22.8S	35.9W
2057 Dec 11	0:52:08	126	P	-0.4853	2.0194	0.9197	359.7	204.8	-	22.6N	14.7W
2058 Jun 06	19:14:17	131	T-	-0.1181	2.6226	1.6628	324.5	214.1	98.0	22.8S	71.1E
2058 Nov 30	3:14:47	136	T+	0.2208	2.4819	1.4277	353.9	221.5	90.4	21.9N	51.6W
2059 May 27	7:54:03	141	P	-0.9098	1.1963	0.1846	282.5	97.9	-	22.2S	119.5W
2059 Nov 19	13:00:04	146	P	0.9004	1.2055	0.2097	271.2	99.9	-	20.4N	161.0E
2060 Apr 15	21:35:33	113	N	1.1622	0.7694	-0.3136	255.7	-	-	9.3S	36.5E
2060 Oct 09	18:52:00	118	N	-1.0671	0.8816	-0.0779	232.0	-	-	5.8N	74.3E
2060 Nov 08	4:02:40	156	Nb	1.5332	0.0286	-0.9356	45.3	-	-	18.2N	65.4W
2061 Apr 04	21:52:32	123	T	0.4300	2.1064	1.0360	355.8	210.3	30.9	5.8S	32.7E
2061 Sep 29	9:36:40	128	T	-0.3810	2.1576	1.1640	325.5	203.1	59.6	2.4N	146.4W
2062 Mar 25	3:32:17	133	T	-0.3150	2.2925	1.2715	339.1	212.0	75.3	2.2S	51.7W
2062 Sep 18	18:32:28	138	T	0.3736	2.1979	1.1515	350.0	213.1	60.2	1.2S	80.2E
2063 Mar 14	16:04:14	143	P	-1.0008	1.0108	0.0363	248.5	41.9	-	1.5N	120.7E
2063 Sep 07	20:39:36	148	N	1.1375	0.8121	-0.2657	261.2	-	-	4.9S	49.1E
2064 Feb 02	21:47:22	115	P	0.9969	1.0215	0.0395	249.8	43.7	-	17.6N	37.0E
2064 Jul 28	7:51:13	120	P	-0.9473	1.1378	0.1055	285.0	76.5	-	19.6S	115.8W
2065 Jan 22	9:57:23	125	T	0.3371	2.2579	1.2248	339.0	209.7	69.4	19.8N	146.3W
2065 Jul 17	17:47:04	130	T-	-0.1402	2.5907	1.6138	331.9	217.0	97.7	21.1S	94.9E
2066 Jan 11	15:03:11	135	T	-0.3687	2.2276	1.1395	361.6	216.0	58.7	21.4N	136.1E
2066 Jul 07	9:28:53	140	P	0.6056	1.7196	0.7770	293.1	172.0	-	21.9S	141.1W
2066 Dec 31	14:28:32	145	N	-1.0540	0.9789	-0.1264	284.2	-	-	22.1N	143.5E
2067 May 28	18:54:30	112	N	-1.2013	0.6416	-0.3316	209.2	-	-	22.8S	75.7E
2067 Jun 27	2:39:28	150	N	1.3394	0.3770	-0.5736	160.5	-	-	22.0S	39.3W
2067 Nov 21	0:03:04	117	N	1.2107	0.6557	-0.3798	222.2	-	-	21.0N	4.5W
2068 May 17	5:40:39	122	P	-0.4852	1.9839	0.9545	337.5	199.8	-	20.0S	86.1W
2068 Nov 09	11:45:21	127	T	0.4645	1.9974	1.0161	312.0	190.9	19.2	17.6N	179.5E
2069 May 06	9:08:18	132	T+	0.2717	2.3977	1.3242	369.1	227.0	85.0	16.5S	137.9W
2069 Oct 30	3:33:26	137	T-	-0.2263	2.4247	1.4628	316.3	206.4	87.4	13.7N	57.4W
2070 Apr 25	9:19:45	142	Nx	1.0044	1.0527	-0.0197	287.7	-	-	12.5S	140.2W
2070 Oct 19	18:49:32	147	P	-0.9406	1.1270	0.1395	263.9	82.3	-	9.4N	74.1E
2071 Mar 16	1:29:28	114	N	-1.0757	0.8890	-0.1183	245.8	-	-	0.8N	20.5W
2071 Sep 09	15:03:59	119	N	1.0835	0.9000	-0.1575	265.9	-	-	4.2S	133.0E
2072 Mar 04	15:21:26	124	T	-0.3431	2.2137	1.2452	313.9	200.1	69.0	5.6N	132.4E
2072 Aug 28	16:04:01	129	T	0.3563	2.2439	1.1673	366.9	221.1	64.8	8.9S	119.2E
2073 Feb 22	7:23:11	134	T	0.3389	2.2230	1.2514	314.6	200.4	69.8	10.3N	107.4W
2073 Aug 17	17:40:59	139	T	-0.3998	2.1490	1.1024	351.3	212.3	50.8	13.4N	95.8E
2074 Feb 11	20:54:15	144	N	1.0612	0.9203	-0.0960	250.2	-	-	14.7N	50.2E
2074 Jul 08	17:19:54	111	N	1.4457	0.1884	-0.7751	117.3	-	-	20.9S	101.3E
2074 Aug 07	1:54:21	149	N	-1.1291	0.7826	-0.2079	232.9	-	-	17.4S	26.9W
2075 Jan 02	9:53:20	116	N	-1.1643	0.7729	-0.3256	255.8	-	-	21.8N	147.3W
2075 Jun 28	9:53:52	121	P	0.6897	1.5639	0.6235	284.2	157.6	-	22.5S	147.6W
2075 Dec 22	8:54:11	126	P	-0.4945	2.0024	0.9028	358.6	203.4	-	23.0N	133.9W

Section 3: Phases of the Moon: 2051 to 2060

Photo 3–1 Various phases of the Moon over one lunar month. ©2010 F. Espenak

Phases of the Moon

Year	New Moon	First Quarter	Full Moon	Last Quarter
2051				Jan 05 04:29
	Jan 12 18:58	Jan 19 12:38	Jan 26 21:20	Feb 04 01:40
	Feb 11 06:41	Feb 17 22:16	Feb 25 14:53	Mar 05 19:47
	Mar 12 16:52	Mar 19 09:34	Mar 27 09:00	Apr 04 09:41
	Apr 11 01:59 P	Apr 17 22:38	Apr 26 02:19 t	May 03 19:30
	May 10 10:29	May 17 13:29	May 25 17:35	Jun 02 02:15
	Jun 08 18:56	Jun 16 05:55	Jun 24 06:14	Jul 01 07:15
	Jul 08 04:09	Jul 15 23:21	Jul 23 16:36	Jul 30 11:52
	Aug 06 15:05	Aug 14 16:49	Aug 22 01:35	Aug 28 17:29
	Sep 05 04:33	Sep 13 09:20	Sep 20 10:11	Sep 27 01:22
	Oct 04 20:47 P	Oct 13 00:12	Oct 19 19:13 t	Oct 26 12:39
	Nov 03 14:59	Nov 11 13:07	Nov 18 05:06	Nov 25 04:02
	Dec 03 09:37	Dec 11 00:07	Dec 17 16:05	Dec 24 23:21

Year	New Moon	First Quarter	Full Moon	Last Quarter
2052	Jan 02 03:05	Jan 09 09:27	Jan 16 04:24	Jan 23 21:03
	Jan 31 18:30	Feb 07 17:35	Feb 14 18:21	Feb 22 18:44
	Mar 01 07:36	Mar 08 01:17	Mar 15 09:54	Mar 23 14:09
	Mar 30 18:27 T	Apr 06 09:28	Apr 14 02:29 n	Apr 22 06:03
	Apr 29 03:20	May 05 19:05	May 13 19:00	May 21 18:15
	May 28 10:50	Jun 04 06:49	Jun 12 10:27	Jun 20 03:10
	Jun 26 17:50	Jul 03 20:59	Jul 12 00:23	Jul 19 09:37
	Jul 26 01:31	Aug 02 13:20	Aug 10 12:52	Aug 17 14:43
	Aug 24 11:07	Sep 01 07:10	Sep 09 00:15	Sep 15 19:48
	Sep 22 23:32 A	Oct 01 01:36	Oct 08 10:54 p	Oct 15 02:22
	Oct 22 15:03	Oct 30 19:39	Nov 06 21:09	Nov 13 11:50
	Nov 21 09:02	Nov 29 12:16	Dec 06 07:18	Dec 13 01:07
	Dec 21 04:15	Dec 29 02:28		

All times are in Universal Time (UT1).
Eclipses at New Moon (solar eclipse) or Full Moon (lunar eclipse), are indicated by these symbols:

Solar Eclipses	Lunar Eclipses
T - Total	t - Total (Umbral)
A - Annular	p - Partial (Umbral)
H - Hybrid	n - Penumbral
P - Partial	

Phases of the Moon

Year	New Moon	First Quarter	Full Moon	Last Quarter
2053			Jan 04 17:46	Jan 11 18:09
	Jan 19 23:12	Jan 27 13:41	Feb 03 04:57	Feb 10 13:49
	Feb 18 16:31	Feb 25 22:09	Mar 04 17:09 n	Mar 12 10:21
	Mar 20 07:11 A	Mar 27 04:50	Apr 03 06:22	Apr 11 06:04
	Apr 18 18:48	Apr 25 11:02	May 02 20:25	May 10 23:40
	May 18 03:43	May 24 18:04	Jun 01 11:02	Jun 09 14:19
	Jun 16 10:51	Jun 23 02:55	Jul 01 02:01	Jul 09 01:47
	Jul 15 17:26	Jul 22 14:16	Jul 30 17:06	Aug 07 10:25
	Aug 14 00:41	Aug 21 04:27	Aug 29 07:53 n	Sep 05 17:05
	Sep 12 09:36 T	Sep 19 21:29	Sep 27 21:50	Oct 04 23:01
	Oct 11 20:53	Oct 19 16:54	Oct 27 10:38	Nov 03 05:37
	Nov 10 10:55	Nov 18 13:26	Nov 25 22:21	Dec 02 14:04
	Dec 10 03:40	Dec 18 09:11	Dec 25 09:23	

Year	New Moon	First Quarter	Full Moon	Last Quarter
2054				Jan 01 01:11
	Jan 08 22:34	Jan 17 02:14	Jan 23 20:08	Jan 30 15:08
	Feb 07 18:14	Feb 15 15:36	Feb 22 06:46 t	Mar 01 07:36
	Mar 09 12:46 P	Mar 17 01:21	Mar 23 17:21	Mar 31 01:50
	Apr 08 04:32	Apr 15 08:23	Apr 22 04:02	Apr 29 20:46
	May 07 17:00	May 14 13:57	May 21 15:16	May 29 15:03
	Jun 06 02:40	Jun 12 19:17	Jun 20 03:42	Jun 28 07:30
	Jul 05 10:34	Jul 12 01:36	Jul 19 17:47	Jul 27 21:28
	Aug 03 17:48 P	Aug 10 10:05	Aug 18 09:22 t	Aug 26 08:56
	Sep 02 01:18 P	Sep 08 21:46	Sep 17 01:41	Sep 24 18:26
	Oct 01 09:49	Oct 08 13:19	Oct 16 17:44	Oct 24 02:39
	Oct 30 20:01	Nov 07 08:34	Nov 15 08:49	Nov 22 10:22
	Nov 29 08:33	Dec 07 06:07	Dec 14 22:41	Dec 21 18:22
	Dec 28 23:52			

Year	New Moon	First Quarter	Full Moon	Last Quarter
2055		Jan 06 03:39	Jan 13 11:21	Jan 20 03:24
	Jan 27 17:39 P	Feb 04 22:59	Feb 11 22:48 t	Feb 18 14:14
	Feb 26 12:39	Mar 06 14:48	Mar 13 08:57	Mar 20 03:18
	Mar 28 07:01	Apr 05 02:43	Apr 11 17:58	Apr 18 18:35
	Apr 26 23:17	May 04 11:10	May 11 02:31	May 18 11:30
	May 26 12:57	Jun 02 17:01	Jun 09 11:36	Jun 17 05:02
	Jun 25 00:15	Jul 01 21:31	Jul 08 22:11	Jul 16 22:15
	Jul 24 09:48 T	Jul 31 02:11	Aug 07 10:57 p	Aug 15 14:26
	Aug 22 18:14	Aug 29 08:35	Sep 06 01:56	Sep 14 05:14
	Sep 21 02:19	Sep 27 18:11	Oct 05 18:38	Oct 13 18:22
	Oct 20 10:49	Oct 27 07:53	Nov 04 12:12	Nov 12 05:38
	Nov 18 20:34	Nov 26 01:42	Dec 04 05:40	Dec 11 15:05
	Dec 18 08:15	Dec 25 22:29		

All times are in Universal Time (UT1).

Eclipses at New Moon (solar eclipse) or Full Moon (lunar eclipse), are indicated by these symbols:

Solar Eclipses	Lunar Eclipses
T – Total	t – Total (Umbral)
A – Annular	p – Partial (Umbral)
H – Hybrid	n – Penumbral
P – Partial	

Phases of the Moon

Year	New Moon	First Quarter	Full Moon	Last Quarter
2056			Jan 02 22:05	Jan 09 23:13
	Jan 16 22:10 A	Jan 24 20:21	Feb 01 12:36 n	Feb 08 07:01
	Feb 15 14:00	Feb 23 17:11	Mar 02 00:40	Mar 08 15:31
	Mar 16 06:52	Mar 24 11:17	Mar 31 10:25	Apr 07 01:33
	Apr 14 23:50	Apr 23 01:33	Apr 29 18:31	May 06 13:30
	May 14 16:06	May 22 11:50	May 29 01:58	Jun 05 03:21
	Jun 13 07:04	Jun 20 18:48	Jun 27 09:47 n	Jul 04 18:55
	Jul 12 20:20 A	Jul 19 23:45	Jul 26 18:54 n	Aug 03 11:52
	Aug 11 07:48	Aug 18 04:13	Aug 25 06:00	Sep 02 05:43
	Sep 09 17:47	Sep 16 09:50	Sep 23 19:34	Oct 01 23:33
	Oct 09 03:00	Oct 15 17:58	Oct 23 11:46	Oct 31 16:12
	Nov 07 12:20	Nov 14 05:33	Nov 22 06:12	Nov 30 06:37
	Dec 06 22:31	Dec 13 20:54	Dec 22 01:34 n	Dec 29 18:22

Year	New Moon	First Quarter	Full Moon	Last Quarter
2057	Jan 05 09:49 T	Jan 12 15:34	Jan 20 20:01	Jan 28 03:44
	Feb 03 22:10	Feb 11 12:25	Feb 19 11:56	Feb 26 11:30
	Mar 05 11:25	Mar 13 09:35	Mar 21 00:45	Mar 27 18:39
	Apr 04 01:31	Apr 12 04:59	Apr 19 10:49	Apr 26 02:06
	May 03 16:32	May 11 21:06	May 18 19:02	May 25 10:40
	Jun 02 08:11	Jun 10 09:30	Jun 17 02:18 p	Jun 23 21:08
	Jul 01 23:47 A	Jul 09 18:37	Jul 16 09:28	Jul 23 10:09
	Jul 31 14:32	Aug 08 01:30	Aug 14 17:21	Aug 22 02:02
	Aug 30 03:54	Sep 06 07:18	Sep 13 02:53	Sep 20 20:25
	Sep 28 16:00	Oct 05 13:13	Oct 12 15:01	Oct 20 16:09
	Oct 28 03:19	Nov 03 20:24	Nov 11 06:24	Nov 19 11:31
	Nov 26 14:22	Dec 03 05:54	Dec 11 00:46 p	Dec 19 05:02
	Dec 26 01:22 T			

Year	New Moon	First Quarter	Full Moon	Last Quarter
2058		Jan 01 18:30	Jan 09 20:39	Jan 17 19:43
	Jan 24 12:14	Jan 31 10:28	Feb 08 15:54	Feb 16 07:16
	Feb 22 22:56	Mar 02 05:10	Mar 10 08:52	Mar 17 15:56
	Mar 24 09:50	Apr 01 01:03	Apr 08 22:55	Apr 15 22:27
	Apr 22 21:29	Apr 30 20:18	May 08 10:12	May 15 03:58
	May 22 10:23 P	May 30 13:33	Jun 06 19:15 t	Jun 13 09:50
	Jun 21 00:35 P	Jun 29 04:13	Jul 06 02:46	Jul 12 17:28
	Jul 20 15:40	Jul 28 16:19	Aug 04 09:37	Aug 11 04:00
	Aug 19 07:03	Aug 27 02:10	Sep 02 16:51	Sep 09 18:07
	Sep 17 22:17	Sep 25 10:14	Oct 02 01:36	Oct 09 11:41
	Oct 17 13:05	Oct 24 17:16	Oct 31 12:54	Nov 08 07:47
	Nov 16 03:09 P	Nov 23 00:16	Nov 30 03:17 t	Dec 08 04:51
	Dec 15 16:12	Dec 22 08:27	Dec 29 20:25	

All times are in Universal Time (UT1).
Eclipses at New Moon (solar eclipse) or Full Moon (lunar eclipse), are indicated by these symbols:

Solar Eclipses	Lunar Eclipses
T - Total	t - Total (Umbral)
A - Annular	p - Partial (Umbral)
H - Hybrid	n - Penumbral
P - Partial	

Phases of the Moon

Year	New Moon	First Quarter	Full Moon	Last Quarter
2059				Jan 07 01:06
	Jan 14 03:57	Jan 20 18:51	Jan 28 15:11	Feb 05 18:49
	Feb 12 14:27	Feb 19 07:57	Feb 27 10:06	Mar 07 08:47
	Mar 14 00:05	Mar 20 23:30	Mar 29 03:47	Apr 05 18:46
	Apr 12 09:28	Apr 19 16:37	Apr 27 19:17	May 05 01:29
	May 11 19:15 T	May 19 10:22	May 27 08:04 p	Jun 03 06:20
	Jun 10 05:57	Jun 18 03:55	Jun 25 18:12	Jul 02 10:53
	Jul 09 17:58	Jul 17 20:34	Jul 25 02:24	Jul 31 16:43
	Aug 08 07:37	Aug 16 11:40	Aug 23 09:42	Aug 30 01:05
	Sep 06 23:01	Sep 15 00:44	Sep 21 17:18	Sep 28 12:54
	Oct 06 15:50	Oct 14 11:38	Oct 21 02:15	Oct 28 04:31
	Nov 05 09:11 A	Nov 12 20:45	Nov 19 13:09 p	Nov 26 23:42
	Dec 05 01:49	Dec 12 04:50	Dec 19 02:11	Dec 26 21:17

Year	New Moon	First Quarter	Full Moon	Last Quarter
2060	Jan 03 16:40	Jan 10 12:52	Jan 17 17:14	Jan 25 19:14
	Feb 02 05:22	Feb 08 21:41	Feb 16 09:56	Feb 24 15:06
	Mar 02 16:11	Mar 09 07:52	Mar 17 03:41	Mar 25 07:08
	Apr 01 01:37	Apr 07 19:42	Apr 15 21:21 n	Apr 23 18:53
	Apr 30 10:11 T	May 07 09:19	May 15 13:39	May 23 03:01
	May 29 18:23	Jun 06 00:44	Jun 14 03:37	Jun 21 08:44
	Jun 28 02:58	Jul 05 17:38	Jul 13 15:08	Jul 20 13:24
	Jul 27 12:49	Aug 04 11:16	Aug 12 00:51	Aug 18 18:23
	Aug 26 00:56	Sep 03 04:36	Sep 10 09:44	Sep 17 01:00
	Sep 24 15:53	Oct 02 20:41	Oct 09 18:41 n	Oct 16 10:30
	Oct 24 09:25 A	Nov 01 10:56	Nov 08 04:17 n	Nov 14 23:48
	Nov 23 04:16	Nov 30 23:11	Dec 07 14:48	Dec 14 17:15
	Dec 22 22:39	Dec 30 09:29		

All times are in Universal Time (UT1).
Eclipses at New Moon (solar eclipse) or Full Moon (lunar eclipse), are indicated by these symbols:

Solar Eclipses	Lunar Eclipses
T - Total	t - Total (Umbral)
A - Annular	p - Partial (Umbral)
H - Hybrid	n - Penumbral
P - Partial	

Photo 3–2 The changing phases of the Moon. ©2010 F. Espenak

The phases of the Moon for the entire 21st Century can be found at the AstroPixels.com website:

www.astropixels.com/ephemeris/moon/phases2001gmt.html

Books by the Author

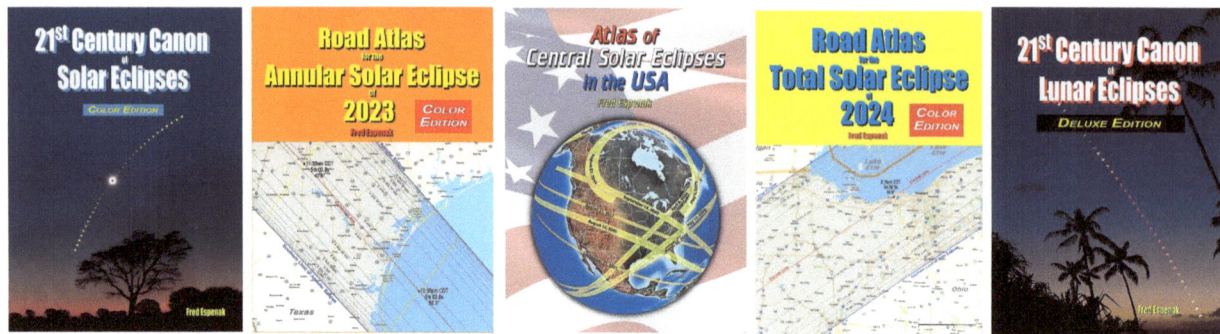

Below is a partial list of some of the books Fred Espenak has written through Astropixels Publishing:

21st Century Canon of Solar Eclipses

The complete guide to every solar eclipse occurring from 2001 tom 2100 (224 eclipses in all). It includes information and maps for all total, annular, hybrid, and partial eclipses. A special world atlas shows detailed full page maps of all central eclipse paths (total, annular and hybrid).

Road Atlas for the Annular Solar Eclipse of 2023

Detailed road maps of the entire eclipse path from the western USA, through Mexico, Central and South America. Information printed on the maps makes it easy to estimate the duration of annularity from any location in the eclipse path. This is the next annular solar eclipse visible from the USA.

Atlas of Central Solar Eclipses in the USA

When was the last total eclipse through the USA and when is the next? How often do they happen? What total eclipse tracks passed across the USA during the 17th, 18th, and 19th centuries, etc., and what states did they include? And how often is a total solar eclipse visible from each of the 50 states? The Atlas of Central Solar Eclipses in the USA answers all of these questions and more with hundreds of maps and tables.

Road Atlas for the Total Solar Eclipse of 2024

This book contains detailed road maps of the entire eclipse path from Mexico, through the USA and Canada. Information printed on the maps makes it easy to estimate the duration of totality from any location in the eclipse path. This is the next total solar eclipse visible from the USA.

21st Century Canon of Lunar Eclipses

The complete guide to every lunar eclipse occurring from 2001 tom 2100 (228 eclipses in all). It includes information and maps for all total, partial, and penumbral eclipses. The predictions use a new model for Earth's elliptical shadows.

For information on these books and more, visit *Astropixels Publishing*:

http://eclipsewise.com/pubs/index.html